Rolls-Royce, Limited

Handbook for 20-25 Rolls-Royce Car

Reprint des Originals von 1933

1. Auflage 2009 | ISBN: 9783941842403

Salzwasser-Verlag (www.salzwasserverlag.de) ist ein Imprint der Europäischer Hochschulverlag GmbH & Co KG, Bremen. (www.eh-verlag.de). Alle Rechte vorbehalten.

Die Deutsche Bibliothek verzeichnet diesen Titel in der Deutschen Nationalbibliografie.

Handbook for 20-25 Rolls-Royce Car

www.salzwasserverlag.de

HANDBOOK
FOR
20-25 H.P.
ROLLS-ROYCE CAR

CHASSIS SERIES
(In order of Issue)

GXO	GOS	GHW	GTZ	GNC	GFE	GOH
GGP	GPS	GRW	GYZ	GRC	GAF	GEH
GDP	GFT	GAW	GBA	GKC	GSF	GBJ
GWP	GBT	GEX	GGA	GED	GRF	GLJ
GLR	GKT	GWX	GHA	GMD	GLG	GCJ
GSR	GAU	GDX	GXB	GYD	GPG	GXK
GTR	GMU	GSY	GUB	GAE	GHG	GBK
GNS	GZU	GLZ	GLB	GWE	GYH	GTK

Liable to alteration without notice

PRICE £1. 5. 0

PUBLISHED BY

ROLLS-ROYCE, LIMITED

Derby, Crewe,
and at 14 and 15, Conduit Street, London

CONDENSED EDITION
Combining NOs. 9, X, XI, XII, XIV, XV, XVI editions

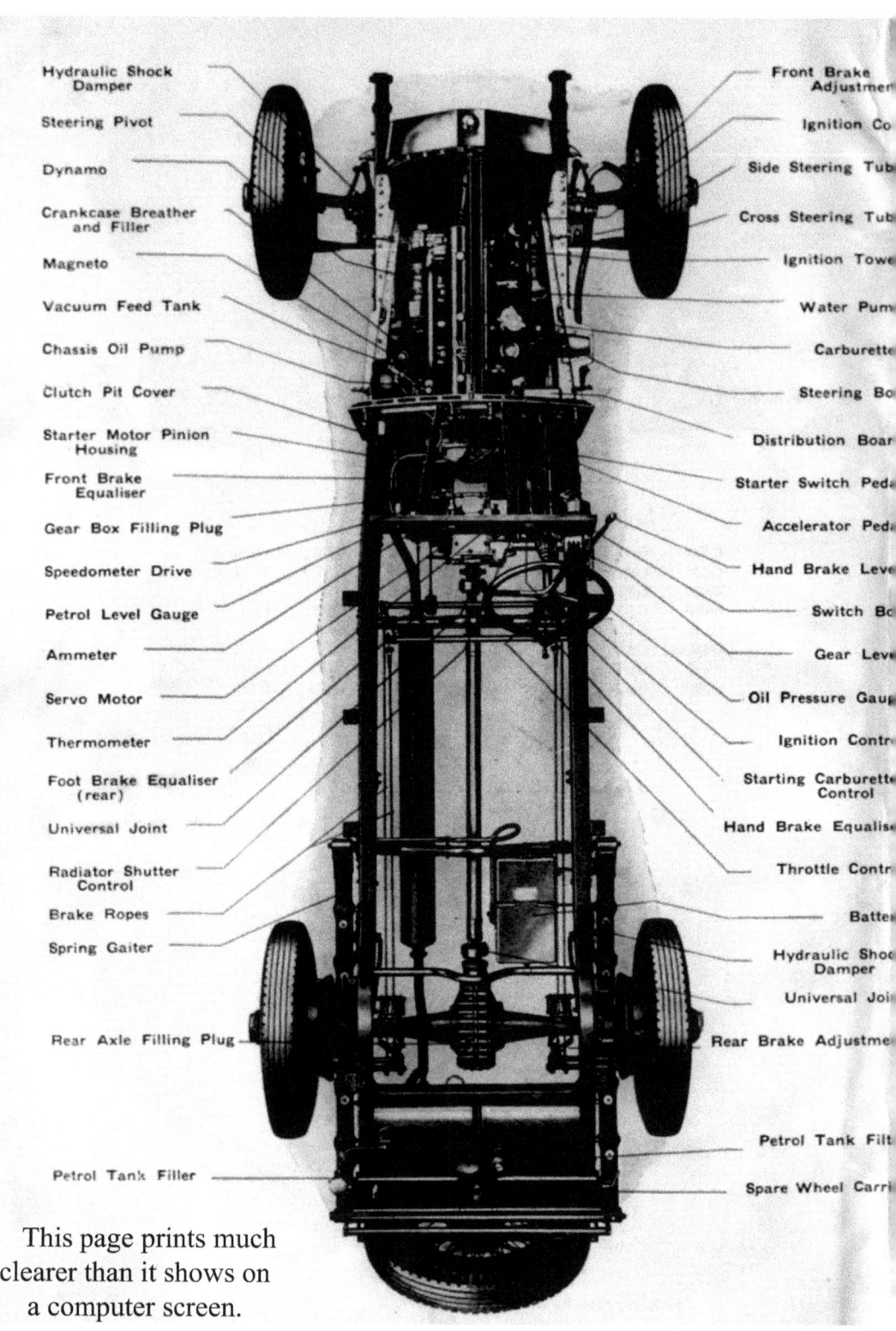

Fig.1. - Plan View of Chassis.

ROLLS-ROYCE LIMITED.

London Office and Showroom :
14 & 15, CONDUIT STREET, LONDON, W.

TELEGRAMS: "ROLHEAD REG. LONDON."
TELEPHONES: MAYfair 6201 (7 lines)
CODES USED: A B C (5TH EDITION), BENTLEY'S, MARCONI
MOTOR TRADE, WESTERN UNION

Main Service Station:
Hythe Road, Willesden, London N.W. 10

TELEGRAMS: "SILVAGOST, WESPHONE, LONDON"
TELEPHONE: LADbroke 2444

Crewe Repair Depôt:
Pym's Lane, Crewe

TELEGRAMS: "ROYCRU, CREWE"
TELEPHONE: CREWE 3271 (10 LINES)

Authorised Paris Service Station:
Franco-Britannic Autos Ltd.
25 Rue Paul Vaillant Couturier, Levallois, Seine

TELEGRAMS: "FRANCOBRIT-LEVALLOIS-PERRET-FRANCE"
TELEPHONE: PERIERE 60-24

THE SECRET OF SUCCESSFUL RUNNING.

Before a Rolls-Royce chassis is sold it is very carefully tested and adjusted by experts. It will run best if no attempt be made to unnecessarily interfere with adjustments.

An owner would do well to instruct his driver as follows:-

Lubricate effectively, in strict accordance with the advice given in this Handbook, and do not neglect *any* part.

Use only those oils which are recommended by Rolls-Royce Limited, who have made prolonged and searching tests on oils. Considerable harm and expense may result from the use of unsuitable oils.

Inspect all parts regularly, but take care not to alter any adjustments unless really necessary.

CONTENTS

	PAGE
SERVICE FACILITIES FOR ROLLS-ROYCE CARS	6
LEADING PARTICULARS OF CHASSIS	7

CHAPTER I.- STARTING THE ENGINE AND DRIVING THE CAR — 9
Starting the Engine - Hand Starting - Ignition Control-Throttle Control - Gear Changing - Battery Charging - Lighting Control and Switch - Radiator Shutters - Controllable Shock Dampers- Over-heating - Fitting of Snow Chains.

CHAPTER II.- PERIODIC LUBRICATION AND ATTENTION ... — 15
Lubricants Recommended-Capacities-Chassis Lubrication System - Drip plugs - Front Axle System - The Oil Gun - Daily Maintenance - Weekly -Every 500 miles - Every 1,000 miles - Every 2,500 miles - Every 5,000 miles - Every 10,000 miles - Every 20,000 miles- Care of Wheels and Hubs-Replacement Tyres -Balancing the Road Wheels.

CHAPTER III.- FUEL SYSTEM ... ... — 37
Fuel Feed - Fuel Filters - The Carburetter (previous to Chassis GYD-25) -Cleaning the Air Valve - Setting of the Jets - Mixture Control - Slow Running - Starting Carburetter - Float Feed Mechanism - Crankcase Breather Pipe to Carburetter - Dismantling the Carburetter - The Carburetter (Chassis Gyd-25 and onwards) - Faulty Adjustment of Carbuetter - Idling and Low-Speed Adjustment - Automatic Air Valve - Float Feed Mechanism.

CHAPTER VI.-ADJUSTMENT OF BRAKES ... ... — 53
General-Adjustment of Rear and Front Brakes (Chassis previous to GAF-1) -Adjustment of Foot and Hand brakes (Chassis GAF-1 and onwards) -Adjustment of Servo.

CHAPTER V.- THE CLUTCH ... ... — 61
Clutch Pedal Adjustment

CHAPTER VI.- COOLANT SYSTEM ... ... — 63
Coolant Pump - CoolantLevel - Frost Precautions.

CHAPTER VII.- ELECTRICAL SYSTEM ... ... — 65
General-Battery-Battery Ignition- Magneto Ignition - Sparking Plugs-Electrical Fault Location

| GUIDE TO MAIN SERVICE STATION ... ... | 72 |
| GIUDE TO CREWE SERVICE STATION ... ... | 73 |

SERVICE FACILITIES
FOR ROLLS-ROYCE CARS

Our interest in your Rolls-Royce car does not cease when you take delivery of the car. It is our ambition that every purchaser of a Rolls-Royce car shall continue to be more than satisfied.

With this end in view, we have appaointed Special Retailers throughout the world, who have established properly equipped Service Stations, staffed by men who have been specially trained in servicing Rolls-Royce cars.

In addition, on the staff of Rolls-Royce Limited, there are experts whose sole duty it is to maintain contact with the Special Retailers, and they are available at all times to be called in for consultation on any matters affecting your car.

If, therefore, you should require any assistance, we ask that you should immediately contact our nearest Special Retailer, who will only be too pleased to place his facilities at your disposal. If necessary he will call in for consultation our expert in that area. It is earnestly hoped that this arrangement will prove of mutual benefit, as we will thus be kept in constant touch with our Customers, who may be spared the trouble of a long journey to one of our Company's Service Stations.

In the event of it being more convenient to call on us direct for assistance, our Main Service Station at Hythe Road, Willesden, London N.W.10, and the one at our factory at Crewe, will be ready at all times to help. (See maps at end of Handbook.)

LEADING PARTICULARS OF CHASSIS.

Engine.
Six cylinders, 3½" (82m/m.) bore, 4½" (114 m/m.) stroke, 3,669 c.c., cubic capacity.

Monobloc with detachable cylinder head, overhead valves operated by pushrods,

Engine Lubrication
Pressure feed to all crankshaft and connecting rod bearings. External oil pump with relief valve, giving a positive low pressure supply to the valve rockers and timing gears.

Carburetter
Rolls-Royce automatic expanding type, controlled by a lever on the steering wheel. Early models are provided with an auxilliary starting carburetter.

Fuel System
Fourteen-gallon or eighteen-gallon tank at rear, depending on date of chassis.

Supply by vacuum feed system with vacuum service tank mounted on the dash-board. Fuel level gauge mounted on the instrument board

Cooling System
By centrifugal pump circulation and fan, with thermostatically-controlled shutters in front of radiator. Hand control for early models.

Coolant temperature thermometer, with warning light mounted on the instrument board on early models.

Electrical Equipment.
Twelve-volt system with automatic regulation of dynamo output. Separate starter motor with Bijur coupling, providing gentle engagement.

Battery of 50 ampere-hour capacity approximately. Twin ignition systems, battery and magneto.

Clutch.
Single dry plate.

Gearbox.
Four forward speeds and reverse, synchromesh or non-synchromesh depending on date of chassis. Right-hand control lever.

Rear Axle.
Full floating type. Spiral bevel drive

Road Springs.
Semi-eliptic front and rear.

Brakes.
Internal expanding, servo operated, on all four wheels. Independent hand brake operating on rear wheels.

Road Wheels.
Dunlop detachable well-base wire wheels, with Dunlop cord, wired type tyres, 6" for 19" rim.

Chassis Lubrication.
Centralised chassis system. Separate axle systems on early models.

Dimensions.

Wheelbase … … … … …	…132" or 129"
Track - Front … … … … …	56 5/16" or 56"
Rear … … … … …	56 5/16" or 56"

CHAPTER I

Starting the Engine and Driving the Car

Starting the Engine - Hand Starting - Ignition Control-Throttle Control - Gear Changing - Battery Charging - Lighting Control and Switch - Radiator Shutters - Controllable Shock Dampers - Overheating - Fitting of Snow Chains.

Starting the Engine.

For chassis series previous to GYD-25, first check that the change gear lever is in neutral, turn fuel tap on dashboard **On**, and close the radiator shutters by moving the control lever on the instrument board: this control is deleted on chassis after GBT-22, as the shutters are thermostatically operated. Next, switch on the ignition by moving the right hand thumb lever on the switchbox to position marked I (Ignition); retard the ignition and close the throttle by bringing both the levers on the steering column to their bottom positions; next open the starting carburetter by pushing the lever on the instrument board to the position marked **Starting** or **On** and set the mixture control lever over to **Strong**. Now depress the small pedal situated low down in the centre of the dashboard; this closes the main switch of the starter motor, and the latter will start up the engine. As soon as the engine commences to run regularly, move the throttle control lever on the steering colimn about half-way up its quadrant and turn back the starting carburetter control lever to the position marked **Running** or **Off**.

The starting carburetter should not be used for more than half a minute before changing over to the main carburetter, and it should only be used when the engine is cold.

Excessive use may lead to failure of the cylinder lubrication owing to dilution of the oil by petrol.

When the engine has been warmed slightly, the mixture control should be set half-way between **Strong** and **Weak**.

For chassis GYD-25 and onwards, first check that the gear lever is in neutral and turn fuel tap on dashboard to M (Main); then switch on the ignition by moving the right-hand thumb lever on the switchbox to position marked I & C, set mixture control thumb lever on dashboard to **Start**, fully retard the ignition and move the hand throttle lever half-way up its quadrant.

Depress the starter button firmly, and to its fullest extent.

As soon as possible after the engine starts, move the mixture cintrol thunb lever to **Normal** or **Run**, and leave it there, and advance the ignition. Nornally the ignition lever should stand about three-quarters along its quadrant.

The mixture control lever is only intended for use when starting from cold; when making a start with a warm engine, leave the thumb lever at **Normal** or **Run**. It is not intended for varying the mixture strength under running conditions.

Difficult starting may be due to dampness in the H.T. distributor caused by condensation. the distributor should be removed under such circumstancesand wiped out with a clean dry rag. The rotor should also be wiped dry. The trouble is only likely to arise when the car has been standing. The warmth of the engine will prevent such condensation normally.

When the engine is cold a high oil presure will be shown on the gauge, due to the greater viscosity of oil at low temperatures. The pressure will fall to the niormal 15 to 20 lbs., as soon as the oil becomes warmer.

Hand Starting.

A starting handle is carried in the tool kit. In the event of it being used, it should be removed afterwards from the bracket and returned to the tool kit.

The ignition must be fully retarded when starting by hand.

Ignition Control.

Under normal circumstances, the ignition lever should be advanced about three-quarters along its quadrant. The actual amount of advance is controlled partly by hand, and partly automatically by means of a centrifugal governor operating on the distributor drive. This is capable of meeting 90 per cent. of the conditions due to varying road speeds, leaving only extreme conditions to be met by moving the hand control on the steering wheel.

Throttle Control.

Under normal running conditions, the hand throttle control should be carried right back in the closed position. An adjustable stop is provided on the carburetter for the throttle lever, which is so adjusted that the engine will idle reliably in these circumstances when the accelerator pedal is released.

On the earlier models, the throttle lever should be set to a position on the quadrant at which the engine will run as slowly as possible without risk of stopping when the clutch is withdrawn.

STARTING THE ENGINE AND DRIVING THE CAR

Gear Changing

The position of the gear lever for each of the four forward speeds and the reverse is shown in Fig. **2**.

On chassis previous to GKT-22, the gearbox does not incorporate synchronous gear meshing devices, therefore it is imperative that on these cars, when making any change of gear, either up or down, that **the double de-clutching method must be used.**

Battery Charging.

For the switchbox with the ignition switch positions marked **I** and **I** & **C**, the following will apply:-

The position marked **I** & **C** (Ignition and Charging) on the switch-box for the thumb lever indicates that the ignition is on, and that the dynamo is charging the battery. This is the position recommended for most running conditions. When, however, the battery is known to be fully charged and the car is running at a moderate speed only, in the

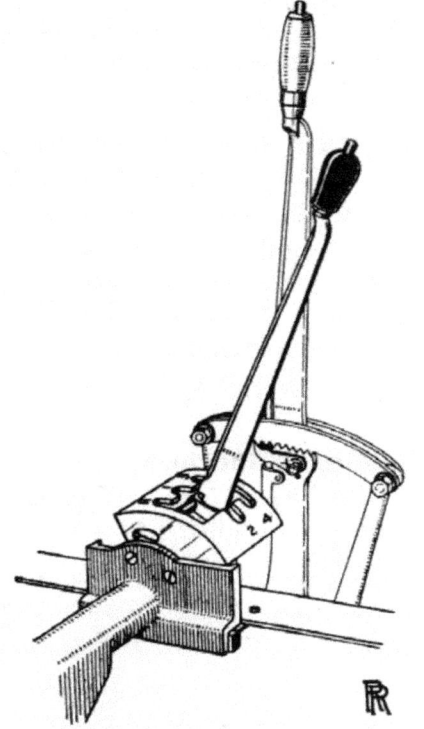

Fig. 2.—PERSPECTIVE VIEW OF GEAR LEVER GATE.

daytime, the charge may be switched off by turning the lever to the position marked I (ignition only).

Whenever the lamps are in use, and the engine is running, always have the switch in the **I** & **C** position.

Where the ignition switch positions are marked **I** & **C** (Summer) and **I** & **C** (Winter) the following will apply:-

With the switch in the **I** & **C** (Summer) position the dynamo output is reduced in order to avoid over-charging the battery, as the demands made upon it in Summer running conditions are usually less than in Winter. In the **I** & **C** (Winter) position the output is increased.

If the battery is known to be in a well-charged condition it is inadvisable to keep the switch a **I** & **C** (Winter) for a long period.

When the head and side lamps are switched on, the dynamo ouput becomes increased, irrespective of the position of the ignition and charge switch.

On chassis GAF-52 and onwards, there is no independent hand control of the battery charge; the provision of an automatic regulator, in combination with a shunt wound dynamo, adjusts the charge rate to suit the state of charge of the battery. When the latter is low, the ammeter on the instrument board will show a higher reading towards **Charge** than it will when the battery is well charged up.

ROLLS-ROYCE 20-25 H.P. CAR

Whenever the switch is at **I & C** and the engine is running above idling speed, the battey is being charged at a rate to suit its state of charge at that particular moment. This can be checked by reference to the ammeter.

Lighting Control and Switch

Adjacent to the ignition is the lighting control switch, for which alternative **On** positions are provided, viz.:-

    OFF ... ... No circuit in action.
    S and **T** ... Side and tail lamps on.
    H, S and **T** ... Head, side and tail lamps on.

This switch may be locked in either the   or   position, and the key withdrawn. Do not attempt to lock the switch in any other position.

Radiator Shutters.

On chassis previous to GBT-22, the radiator shutters are hand-operated by means of a lever on the left-hand side of the instrument board.

Fig. 3.- THERMOSTAT CONTROL OF RADIATOR SHUTTERS.
    L. Coolant control   R. Spring-loaded Pin.
    T. Thermostat   R1. Shutter Lever

STARTING THE ENGINE AND DRIVING THE CAR 13

A thermometer arranged on the instrument board indicates the coolant temperature of the engine, a small red lamp warns the driver when the temperature conditions of the engine require adjustment.

The normal working temperature should be between 70° C. and 90° C., and therefore, when starting the engine, the shutters should be closed. They should remain so until the water temperature reaches 70° C.

On chassis GBT-22 and onwards, the radiator shutters are controlled automatically by means of a thermostat in the upper radiator tank.

The thermostat commences to open the shutters when the coolant reaches a temperature of about 60° C., and they are fully open at about 90° C. Under normal running and atmospheric conditions the coolant temperature is maintained between 70° C and 75° C.

A thermostat is provided on the instrument board to indicate that the shutters are operating properly and that there is no shortage of coolant.

Provision has been made for quick disconnection of the shutters in case of defective operation of the thermostat. The spring loaded pin **R** (Fig. 3) should be raised, and the end of the lever **R1** disengaged from the thermostat rod. The shutters should then be left wide open.

Controllable Shock Dampers.

Fitted to chassis GYD-25 and onwards, these provide comfortable riding at all speeds, a centrifugally-controlled pump is incorporated which causes the damper loadings to increase with the road speed. In addition, there is a lever above the steering wheel, marked **Riding Control**, the effect of which is superimposed on that of the governor.

For ordinary town work or touring with moderate speeds, it will be found that the damper loadings as set by the governor are adequate when the hand lever is at either **Minimum** or mid-way. On the other hand, at high speeds or with heavy loads, improved riding comfort will be obtained by moving the lever to **Maximum**.

Overheating.

If on long ascents which call for full throttle, "boiling" should occur due to abnormal conditions of atmospheric temperature, and, or, following winds, etc., it is preferable to change into a lower gear and reduce the throttle opening.

Adjustment of the fan belt may be necessary, and this should receive attention.

Fitting of Snow Chains.

In the event of snow chains being necessary, they should be fitted to the rear wheels only.

A Parsons' chain, known as the "Special Rolls-Royce Type", is available. It is recommended that these be obtained through Messrs. Rolls-Royce Ltd., or one of their "Special Retailers", in order to ensure the supply of the correct type.

When fitting these special chains, it is *essential* to commence by fastening the one hook on the inside of the wheel and always to take up the adjustment on the outside, where two fastening clips are provided. The tensioning springs which are supplied to go on the outside of the wheel must always be fitted.

CHAPTER II

Periodic Lubrication and Attention

LUBRICANTS RECOMMENDED

Rolls-Royce Limited recommend a first quality oil of viscosity S.A.E. 30 for the engine and gearbox all year round.

NOTE.- The recommended oil for use in the engine may require slight modification if the engine is in very poor condition, in which case use of an S.A.E. 40 viscosity oil may be to advantage.

Any of the following oils are suitable:-

	"A" Engine. S.A.E. 30.	"B" Gearbox S.A.E. 40.	"C"	Hand-oiling Points.
B.P. Energol	S.A.E. 30	S.A.E. 40	S.A.E. 30	S.A.E. 20
Wakefield's Castrol	X.L.	X.X.L.	X.L.	Castrolite
Shell	X. 100	X.100	X. 100	X. 100
	S.A.E. 30	S.A.E. 40	S.A.E. 30	S.A.E. 20
Vacuum "Mobiloil"	A.	BB.	A.	Arctic.

Equivalent oils to the above are also marketed by Sternol Ltd., Alexander Duckham & Co. Ltd., Esso Petroleum Co. Ltd., Gulf Oil (Great Britain) Ltd., and dalton & Co. Ltd.

In the instructions which follow, reference is made to Oil "A", "B" or "C".

Rear Axle.

Oil of viscosity S.A.E. 80/90 as under:-

B.P.	..	..	..	Energol S.A.E. 90.
Wakefield's	..	..	..	Castrol S.T.
Shell	..	..	..	Dentax 90.
Vacuum	..	..	..	Mobilube C.90.

Steering Box.

Oil of viscosity S.A.E. 40 as under "A" above.

Chassis Oil Pump..

Oil of viscosity S.A.E. 30 as under "A" above.

Shock Dampers and Distributor Lubricator.

As under "C" above.

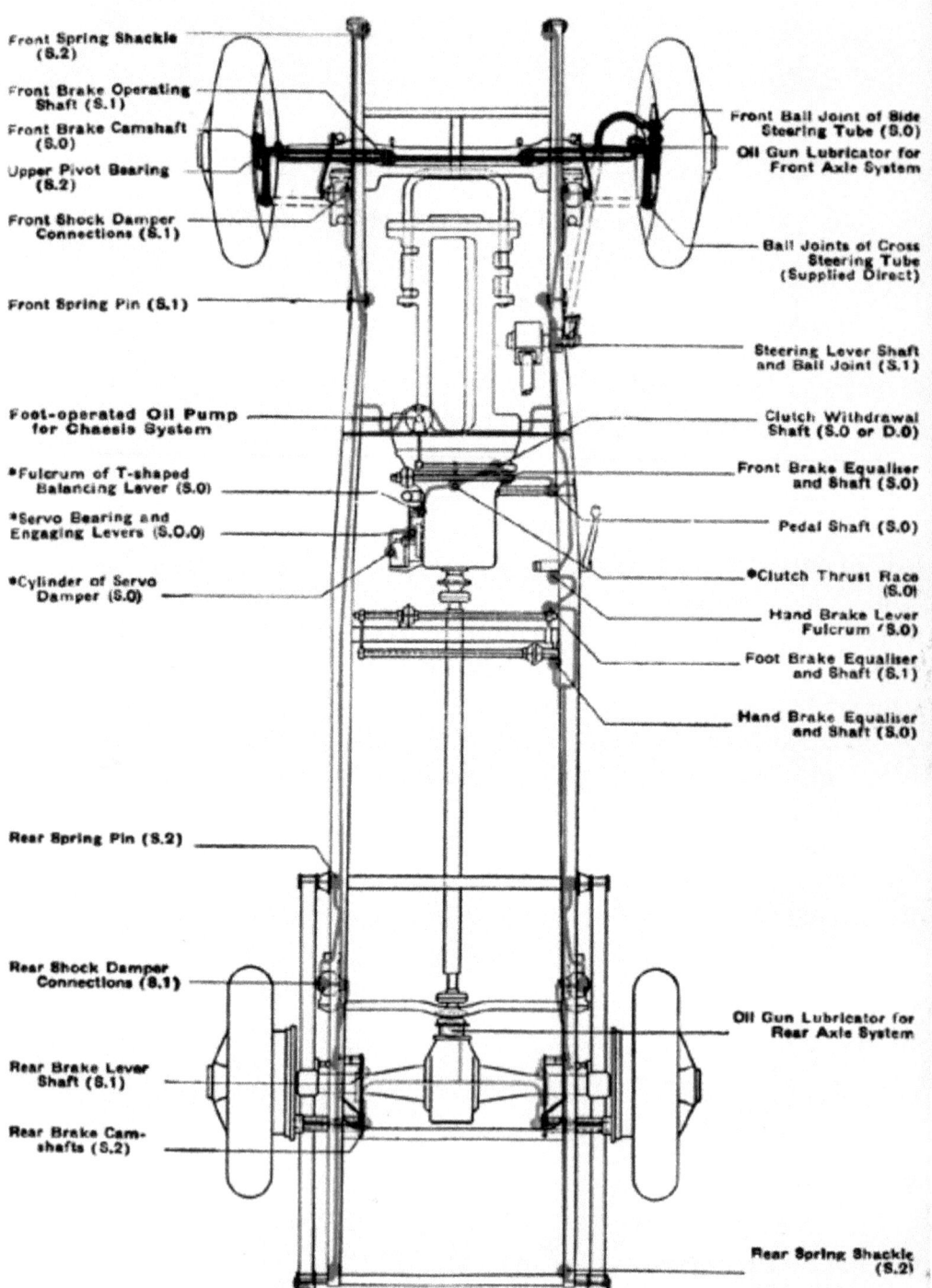

Fig. 4.—DIAGRAM OF CHASSIS LUBRICATION SYSTEM.

(CHASSIS NOS. GXO-II-III, GGP, GDP, GWP, GLR, GSR, GTR, GNS, GOS, GPS, GFT, GBT.)
Frame System coloured RED.   Front Axle System coloured BLUE.   Rear Axle System coloured GREEN.

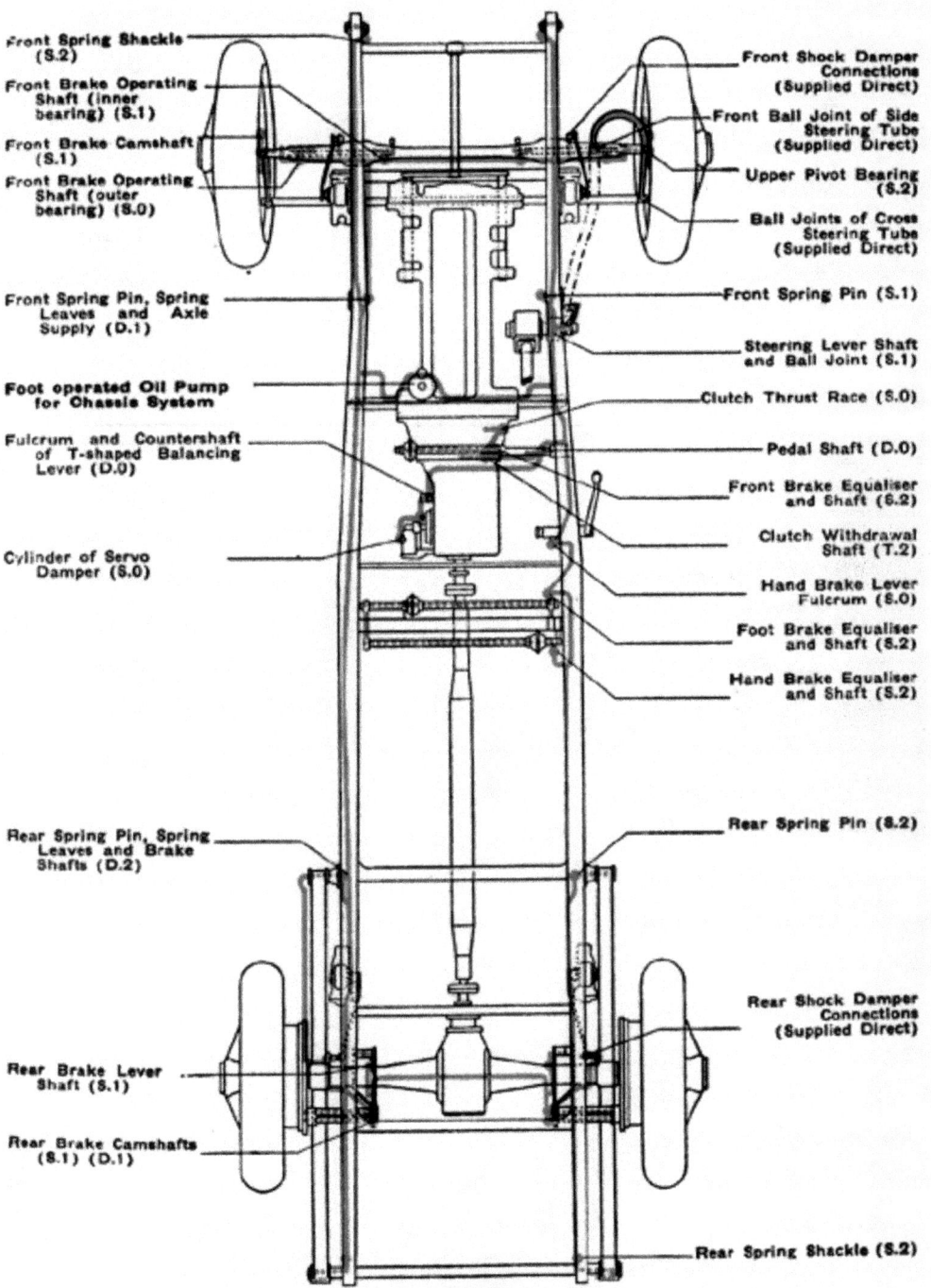

Fig. 5.—DIAGRAM OF CHASSIS LUBRICATION SYSTEM.

(CHASSIS NOS. GKT, GAU, GMU, GZU, GHW, GRW, GAW, GEX, GWX, GDX, GSY and GLZ.)

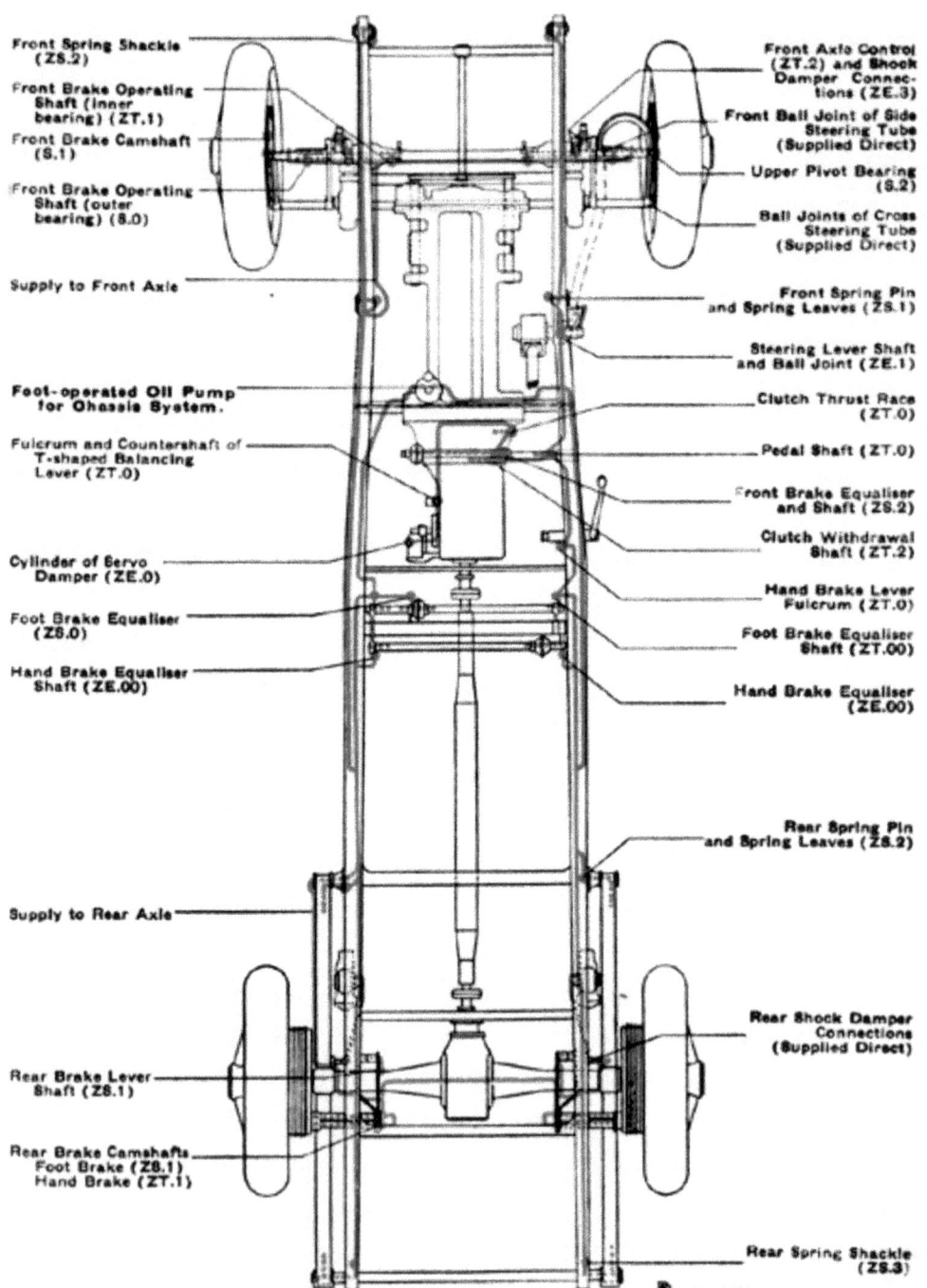

Fig. 6—DIAGRAM OF CHASSIS LUBRICATION SYSTEM.

(CHASSIS NOS. GTZ, GYZ, GBA, GGA, GHA, GXB, GUB, GLB, GNC, GRC, GKB, GED, GMD, GYD, GAE, GWE, GFE, GAE, GSF, GRF, GLG, GPC, GHG, GYH, GOH, GEH, GBJ, GLJ, GCJ, GXK, GBK and GTK.)

(Figs. 4, 5 and 6 print much clearer than they show on a computer screen.)

Front and Rear Hubs.

Belmoline "C", Retinax "R.B"., or similar type of ball-bearing grease.

Propeller Shaft, Contact Breaker Cam and Wheel Hub Shells.

Retinax C.D., Mobilgrease No. 2, or a similar type of grease.

NOTE.- For the PROPELLER SHAFT of all chassis previous to GKC-22 use an oil of viscosity S.A.E. 80/90 as for "REAR AXLE" above.

Water Pump

Belmoline A, Retinax P, or a similar type of grease.

CAPACITIES

Engine	1¼ gallons
Gearbox	5 pints
Rear Axle	2 pints
Chassis Oil Pump	pints
Cooling System	3¾ gallons
Fuel Tank	gallons

CENTRALISED CHASSIS LUBRICATION SYSTEM

General

A foot-operated pump, with which is combined an oil reservoir, is located on the front of the dashboard, and supplies oil under pressure for chassis lubrication.

A diagram of the complete systems, Fig. **4, 5** and **6** with their relative chassis series, show the piping being coloured red with red discs to indicate the position of the drip plugs. The rating of the latter is given in parentheses against the description of the part lubricated.

It should be noted that Fig. **4** shows the Front and Rear Axle Systems coloured in Blue and Green, and that in the chassis series to which this applies these systems must be lubricated by means of the hand-operated Oil Gun.

Foot-operated Oil Pump.

The chassis oil pump is shown in Fig **7**. Normally no attention to the system is necessary beyond filling of the reservoir with oil "C" after removal of the filling plug **A**. This should be done every 2,500 miles, as directed on page 22.

When the reservoir is nearly empty it will be found that the pedal returns instantly after depression, due to the presence of air in the system.

PERIODIC LUBRICATION AND ATTENTION.

On the other hand, if the pedal takes an abnormal length of time to return to its raised position, this may indicate that the felt strainer located at the botom of the reservoir is choked.

Under these circumstances, the cap nut **B** should be unscrewed and the cover **C** removed. The felt strainer pad and its perforated backing plate can be taken out. The pad should be discarded and a new one fitted.

When replacing the parts the perforated plate should first be fitted on the stud, followed by the felt pad and then the distance piece. Care must be taken to see that the vellumoid packing washer is in position between the cover **C** and the casing, and also that the aluminium washer between the nut and the cover is replaced before tightening the nut.

Fig. 7. - FOOT-OPERATED CHASSIS OIL PUMP
A. Filler Plug   C. Strainer Cover
B. Cap Nut

Normally, the felt strainer pad should be discarded and a new one fitted every 20,000 miles, as directed on page 24.

Under no circumstances must any attempt be made to further to dismantle the pump. If any defect in operation should develop, which is not rectified by renewing the strainer pad as directed, the whole unit should be removed from the dashboard and returned to Rolls-Royce Ltd. for correction.

Drip Plugs.

The drip plugs are non-adjustable and non-demountable, and are lettered and numbered to indicate their shapes and relative rates of oil emission respectively, a higher number indicating a greater rate.

The drip plugs never require cleaning, and, being non-demountable no attempt must be made to take them apart. If one is suspected of being defective, it should be replaced with a new plug of the same rating.

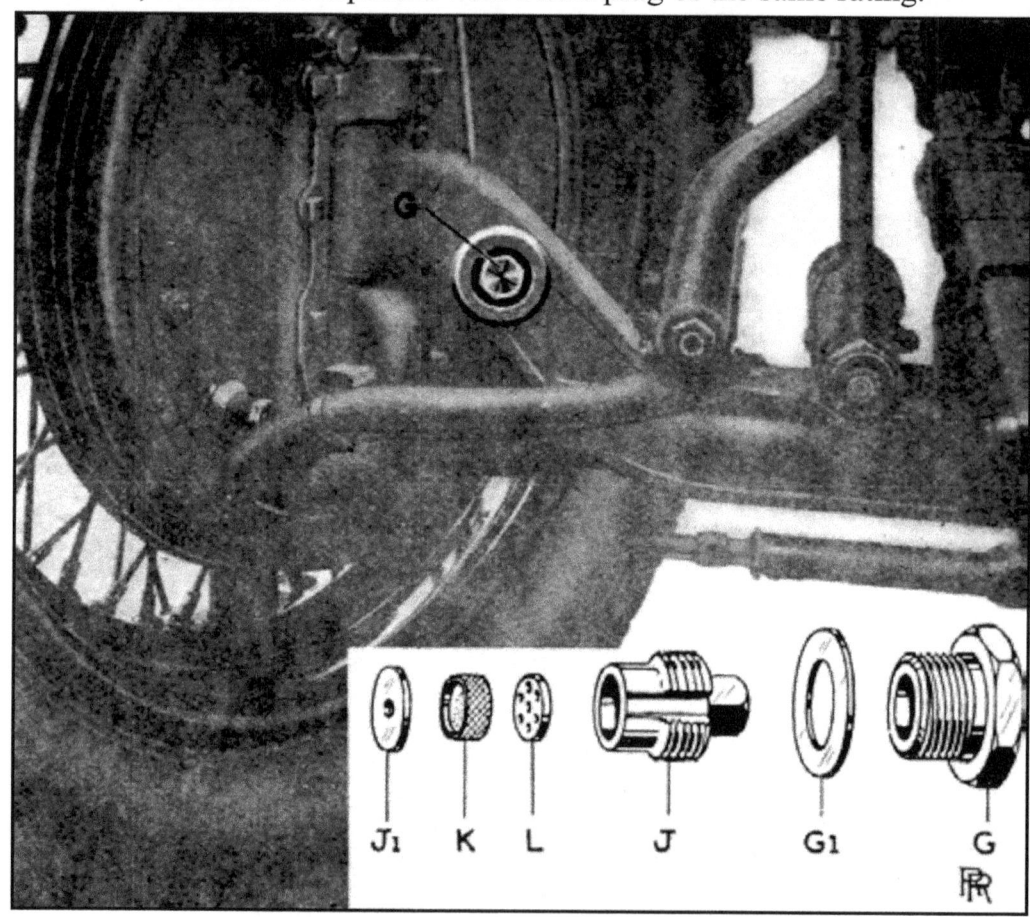

Fig. 8. - STRAINERS ON FRONT AXLE
G. Outer Plug.   L. Backing Washer
G1. Joint Washer.   K. Felt Strainer.
J. Inner Plug   J1. Packing Washer

Front Axle System.

The arrangement of the front axle system renders it necessary to provide separate strainers. For convenience these are located on the ends of the axle, as shown at **G** above, which also shows the component parts inset.

PERIODIC LUBRICATION AND ATTENTION.

The felt strainers should be renewed every 20,000 miles, as directed on page 30, the procedure being as follows:-
1. Carefully clean the outsides of the fittings with a brush and paraffin to prevent the ingress of dirt during dismantling.
2. Unscrew the outer plug, **G**, with a box spanner.
3. Unscrew the inner plug, **J**, with a box spanner. This plug carries the felt strainer, **K**, which should be removed and discarded.

All parts should be carefully cleaned and freed of every trace of grit before replacing. The perforated backing washer, **L**, must be replaced in the inner plug, **J**, before fitting the new felt strainer, **K**, with its gauze-covered side towards the washer.

Two aluminium packing washers are provided, one, **J**$_1$, between the inner plug, **J**, and the bottom of the recess in the axle, and the other, **G**$_1$, under the shoulder of the outer plug, **G**. Care must be taken to replace these.

LUBRICATION BY MEANS OF THE OIL GUN

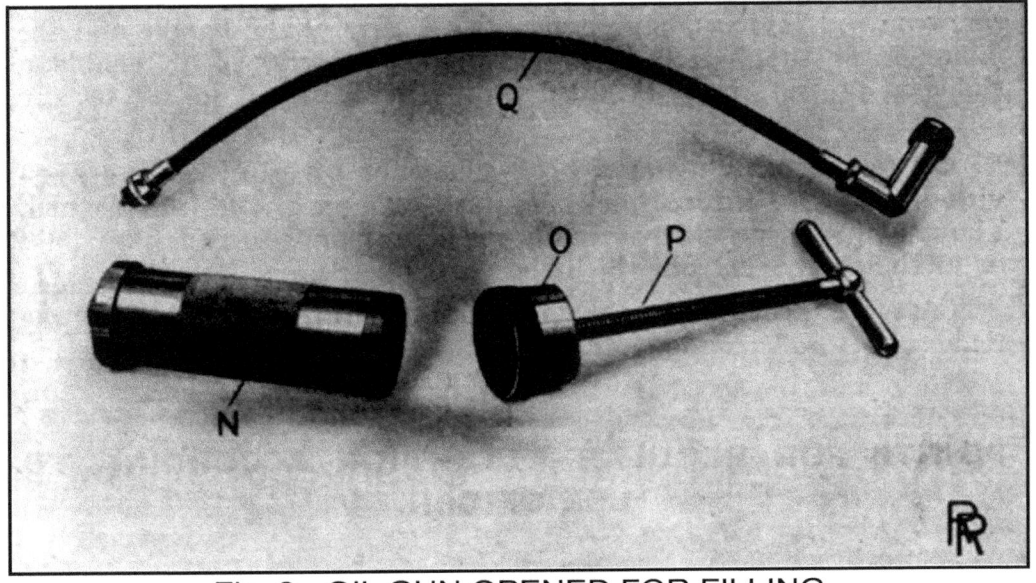

Fig. 9.- OIL GUN OPENED FOR FILLING
N. Barrel.   P. Piston Rod.
O. Cap.   Q. Connection.

The oil gun is shown dismantled for filling in Fig. **9**. It consists of a barrel **N** on to which is screwed the cap **O**. The rod **P**, carrying a cup leather, is threaded into the cap, therefore when this rod is screwed down by means of the handle, oil may be expelled from the barrel under considerable pressure.

PERIODIC LUBRICATION AND ATTENTION.

The flexible connection **Q** is fitted with a non-return valve, which is closed with a spring, except when the connection is scewed on to one of the chassis adapters or lubricators. Consequently, no oil can be expelled through the connection until this is in position on the lubricator.

In addition, each lubricator on the chassis is equippped with a ball non-return valve, which is opened by the valve in the flexible connection when it is screwed on.

Only the recommended oils should be used in the oil gun.

The gun is filled by unscrewing and removing cap **O**, together with rod **P** and pouring oil into the barrell. To facilitate re-entry of the cup leather into the barrel, the cap **O** is formed with an internal diameter equal to that of the barrel, and before replacing the cap it should be screwed down on the rod as far as possible, as shown in Fig. **9**. The leather will then be contracted by the cap, and on replacement of the latter will enter the barrel freely. The gun is then ready for use.

Owing to the arrangement of the valve in connection **Q**, care must be taken that this is screwed well home on the lubricator, otherwise the gun will not work.

The oil gun is of a special low pressure type, the angle of the screw on rod **P**, in combination with the size of the handle and the diameter of the barrel, being carefully proportioned to enable a sufficient pressure to be attained without undue effort for use on any lubricator on the chassis.

On no account must a high-pressure oil gun, or one provided with an "intensifier", be used on any of the lubricators. The use of such a gun may easily result in damage to the pipe lines or to the component on which the gun is used.

Caps are provided on the oil gun lubricators, which must be removed before screwing on the connection and subsequently replaced.

Points of Regular Attention according to use of Car

DAILY.

Crankcase Oil.

The engine oil level indicator situated on the left-hand side of the crankcase should be inspected *daily*, and the quantity of oil maintained at about one gallon and a quarter, as shown by the indicator finger. The engine should never be run with less than three-quarters of a gallon of oil. The oil filler is on the left-hand side of the engine, the cap being provided with a bayonet joint.

PERIODIC LUBRICATION AND ATTENTION.

Water in Radiator.
The radiator water level should be inspected *daily*. It should be maintained at about half-way across the upper radiator water pipe. (See Fig. **3**.)

Chassis Lubrication.
Depress oil pump pedal *once* while engine is being started for first time in the day, and subsequently *once* every 100 miles. Use the pump more frequently during bad weather.

WEEKLY

Tyres.
Check the tyre pressures. These should be:-
- Front ... ... 35 lbs. ⎫
- Rear ... ... 35 lbs. ⎭ Cold

EVERY 500 MILES.

Front Axle System. (See Fig.**4**.)
Oil should be injected by means of the oil gun into the lubricator situated towards the off-side end of the axle, the handle being turned until it becomes tight. The axle system is then charged.

EVERY 1,000 MILES.

Rear Axle System.
Oil should be injected by means of the oil gun into the lubricator situated on the rear axle casing, access to which is obtained by removing the rear floor boards.

Turn the oil gun handle until it becomes tight, the axle system is then charged.

Battery.
Inspect the level of the acid in the cells, and if necessary, top up with distilled water so as to maintain the level at ½" above the tops of the plates.

EVERY 2,500 MILES

Starter Motor, Dynamo Bearings and Dynamo Coupling.
Inject two or three drops only of oil "C" with the oil can into each lubricator - one on the starter motor and two on the dynamo and oil hole in coupling.

Front Engine Support.
Inject oil "B" using oil gun, and screw down until oil exudes from the ends of bearing (one lubricator).

Carburetter (Chassis previous to GYD-25).
Remove and clean air valve and chamber. Use no lubricants on these parts.

Battery Ignition Governor.

Inject a few drops of oil "C" with the oil can into spring lid lubricator.

Cam of Battery Ignition Contact Breaker.

Smear a trace only of grease on cam surface.

Water Pump Bearing and Gland.

Remove lubricator cap, fill one-third full of recommended grease and screw right down, preferably when engine is warm.

Steering Box.

Remove plug and fill with oil of viscosity S.A.E. 40 as under "A" to mouth of plug orifice.

Starter motor Switch (Oil immersed type).

Remove filler plug and oil "C" if required; the switch should be kept full of oil.

Chassis Lubrication.

Inspect oil level in reservoir and add more oil if necessary.

Brake Connections. (Chassis Nos. see Fig. 4).

Lubricate with oil "C" by means of the oil can the points listed, the figures indicating the number of points requiring attention:-

> Fulcrum of brake actuating levers on the servo shaft (2).
> Servo shaft (1) oil hole.
> Servo engaging levers (1) - one or two drops in oil hole.
> Fulcrum of balancing lever (1).

Propeller Shaft Sliding Joint.

Remove plug and inject about one tablespoonful of oil "C", using the oil can.

Spring Gaiters (where applicable).

Lubricate by means of the oil gun using oil "C"; screw down three or four turns on each of the twelve lubricators.

Fan

Check the fan belt adjustment, and, if necessary, adjust so that at a point equidistant between the pulleys, the fan belt can be moved transversely with the fingers through the distance of ¾".

PERIODIC LUBRICATION AND ATTENTION.

Fig. 10.- FAN BELT ADJUSTMENT.
A. Lubricator   C. Adjusting nut.
B. Lock Nut.

Brakes.
Check and adjust if necessary.

Wheels
Test hub nuts for tightness with the spanner.

Valve Rocker Clearances.
Check and reset the valve rocker clearances if necessary; they should be .004" *with the engine cold.*

The method of adjusting the valve rocker clearances is illustrated in Fig.11.

Fig. 11.- ADJUSTING THE TAPPETS.
U. Tappet head.   W. Feeler Gauge
V. Lock Nut

Remove the valve rocker cover and the two covers on the left-hand side of the engine as shown, taking care that when removing breather pipe from valve rocker cover to air silencer, the fingers only should be used when compressing its two ends against the spring.

The valve tappets are provided with adjustable heads, the tappet head **U** being screwed into the tappet and locked with a nut **V**. On releasing this nut the tappet can be screwed in or out as may be required.

With the engine cold and the valve roller on the base of the cam, a feeler gauge should be inserted as shown at **W**.

Before commencing to adjust a tappet, it should be ascertained that that particular tappet roller is well away from the cam, which is best done by turning the crankshaft by hand until the valve has opened and closed, and then cranking round half a revolution beyond this point.

As each tappet is adjusted, its locknut should be securely tightened up.

PERIODIC LUBRICATION AND ATTENTION.

EVERY 5,000 MILES

Carburetter (Chassis previous to GYD-25).
Remove cover and wipe out interior of float chamber. (See page 46).

Guide of Carburetter Air valve (*Chassis GYD-25 and onwards.*)
Inject one or two drops of oil "C" into small lubricator on side of carburetter. *Be certain afterwards to close the lubricator. (See page 52.)*

Fig. 12.- OIL PLUGS IN REAR AXLE CASING.
A. Stand Pipe
B. Level Plug and Drain.
Q. Filler Plug

Clutch Shaft and Levers.

Remove Clutch pit cover, turn withdrawal sleeve with fingers until slot is at the top, then turn crankshaft until oil hole is visible. Inject a few drops of oil "C". Excess of oil will cause clutch trouble.

Also, lubricate fulcrum pins of clutch levers.

Servo Bearing.

Inject one or two drops of oil "C" with the oil can into spring lid lubricator.

Gearbox

Inspect oil level by means of dip stick when gearbox is warm. Level should stand at notch in flat of stick. Add oil "B" if necessary.

Fig. 13.- OIL PLUGS IN REAR AXLE CASING.
A. Level Plug
Q. Filler Plug
B. Drain Plug.

PERIODIC LUBRICATION AND ATTENTION. 27

Rear Axle

Inspect oil level when axle is warm, and top up, if necessary with the recommended oil, through the filler plug **Q** at the top of the casing.

On chassis previous to GAU-1, the plug at the bottom of the casing (see Fig. **12**) communicates with the interior through a standpipe which projects inside the casing to act as an oil level indicator. This plug should be removed for testing the oil level, and one should not be deceived by the appearance of a small quantity of oil, which is possibly onlt what has lodged in the standpipe.

Chassis GAU-1 and onwards, the stanspipe level is replaced by a level plug on th side of the casing (see Fig. **13**) and oil should be poured into the filler plug hole until it just commences to run out of the level plug hole.

Fan

Inject a few drops of oil "C" into lubricator (**A** Fig. **10**).

Contact Breaker of Battery Ignition

Move aside spring which retains rocker arm and lubricate pivot pin with one drop of oil "C".

Bonnet Ventilators, Fasteners and Locks

Carefully lubricate with the oil can to avoid squeaks and rattles.

Brake Connections etc.

Lubricate with oil "C" by means of the oil can, the points listed hereunder, the figures indicating the number of points requiring attention:-

      Jaws of brake ropes front and rear (12)
      Brake connections and adjustments on rear axle (12)
      Ball joints of front brake pull rods (4). (Remove leather
          stockings).
      Front brake adjustments (4).
      Jaws of brake rods between balancing lever and equalisers
          (front and rear) (4).
      Jaws of brake rod from servo to equaliser (2).
      Joints of coupling rods from servo to balancing lever (4).
      Joints of links between cross shaft and servo (2).
      Jaws of rod from pedal to cross shaft (2).
      Clutch pedal connection jaws (2).
      Accelerator pedal bearing (1).
      Jaws of rod from hand brake lever to equaliser (2).
      Hand brake pawl connections (4).
      Reverse catch of gear lever (3).

Control Mechanism.

On steering wheel, steering column, engine, carburetter, ignition tower and magneto; also radiatot shutter control, shock damper controls. Apply a drop of oil "C" with the oil-can to each bearing and joint.

Engine Oil Strainer

When engine is warm, drain crankcase and remove and clean crankcase oil strainer. Refill with fresh oil to correct level.

Fuel Filters.

The petrol filter fitted on the front of the dashboard should be dismantled and cleaned, also those in the fuel tank, if fitted. (See Figs. 19 and 21).

The filter fitted in the supply pipe to the vacuum tank should also be removed and cleaned. (See page 37).

Similarly, if fitted, the filter arranged in the fuel inlet to the carburetter float chamber should also be cleaned. (See Fig. 20.)

Air Cleaner.

Remove cleaner element and carefully wash in petrol or paraffin. When touring on the Continent this should be done every 2,500 miles.

Fuel Tank.

Release (but do not *remove*) drain plug at bottom of main tank to allow any accumulated water to escape

Spark Plugs.

Remove and clean. Set gaps to .020".

Contact Breaker.

Inspect L.T. make and break contacts of battery ignition.

Set gaps to .017" to .021".

Wheels.

Remove , grease interiors and hubs, and replace.

Steering Joints.

Test steering joints and shock damper connections for play and adjust if necessary.

EVERY 10,000 MILES

Hydraulic Shock Dampers.

Inspect oil level and add more oil if necessary, using the small syringe. *Use only correct oil.* (See page 15.)

PERIODIC LUBRICATION AND ATTENTION.

Shock Damper Control.
Inspect oil level by removing plug from filler spout. *Use only correct oil.* (See page 15.)

Dynamo.
Remove cover, clean away brush dust and inspect brushes.

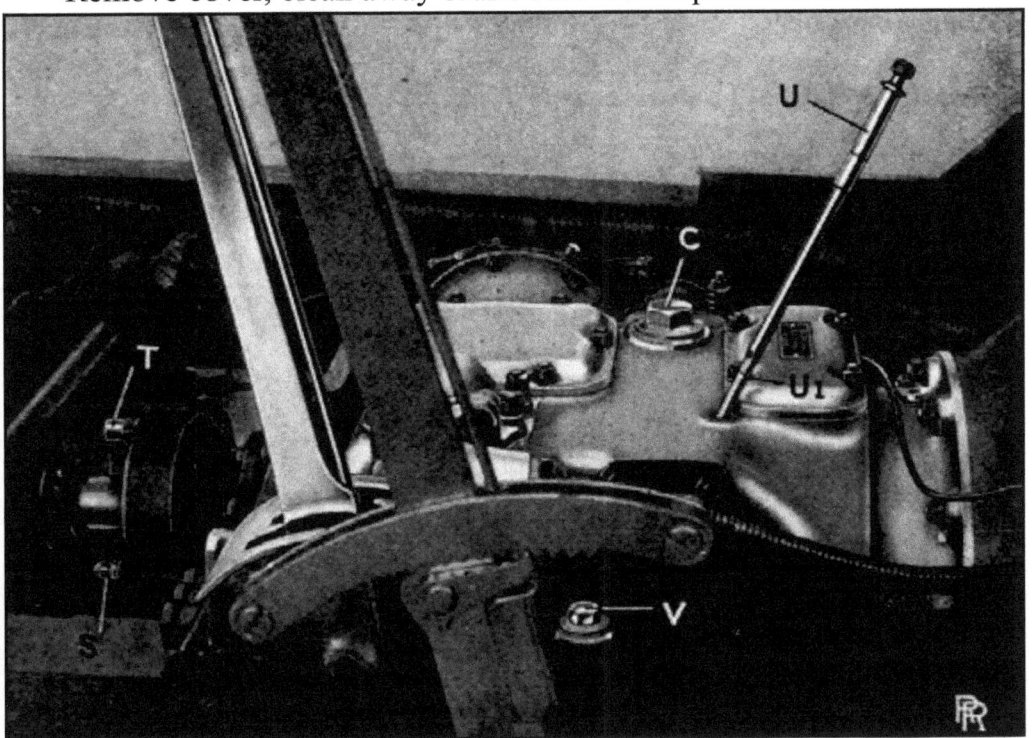

Fig. 14.- GEARBOX AND FORWARD UNIVERSAL JOINT

Universal Joints.
Turn propeller shaft so that lubricators **S** (Fig. 14) are at bottom and air release plugs **T** at top. Remove plugs then inject grease with gun until it commences to flow freely from plug hole. Carefully replace plugs. *On no account operate gun with vent plugs in position.*

Crankcase Btreather Pipe to Carburetter.
Remove and clean gauze between pipe flange and carburetter. (See page 47).

EVERY 20,000 MILES
Gearbox and Rear Axle.
Drain out all the oil when warm by removing drain plugs and filler plugs. (There are *two* drain plugs in gearbox.)

On chassis previous to GAU-1, remove stand pipe from the rear axle to drain (See Fig. 12).

Refill with fresh oil to correct level.

Use only correct oil for cleaning out casings. Do not use petrol, paraffin or other oil solvents.

Chassis lubrication System.

Remove and discard three felt strainer pads located, respectively one at base of chassis oil pump and one at each end of front axle. Replace with new pads.

Brake Servo.

Test adjustment and readjust if necessary.

Air Cleaner.

Renew cleaner element.

CARE OF DUNLOP WHEELS AND HUBS.

Fig. 15.- REMOVING DETACHABLE WHEEL.

Removal of Wheel.

Dunlop detachable wire wheels are fitted, and a special spanner is provided in the tool-kit for removing and replacing them.

In Fig. **15** the spanner is shown in postion on a wheel.

Before using the spanner, the central screw **P** must be unscrewed as far as possible. After jacking up the car, the spanner can be placed in position by pressing the levers **Q** to clear the shoulder on the hub

PERIODIC LUBRICATION AND ATTENTION.

nut. On releasing these levers, it should be noticed that they fit correctly into the groove provided for the purpose.

Screw **P** should then be turned until the serrations of the locking plate **R**, are seen to be clear of those on the hub nut. The latter can then be turned in an anti-clockwise direction and the wheel withdrawn, the hub nut remaining in the spanner.

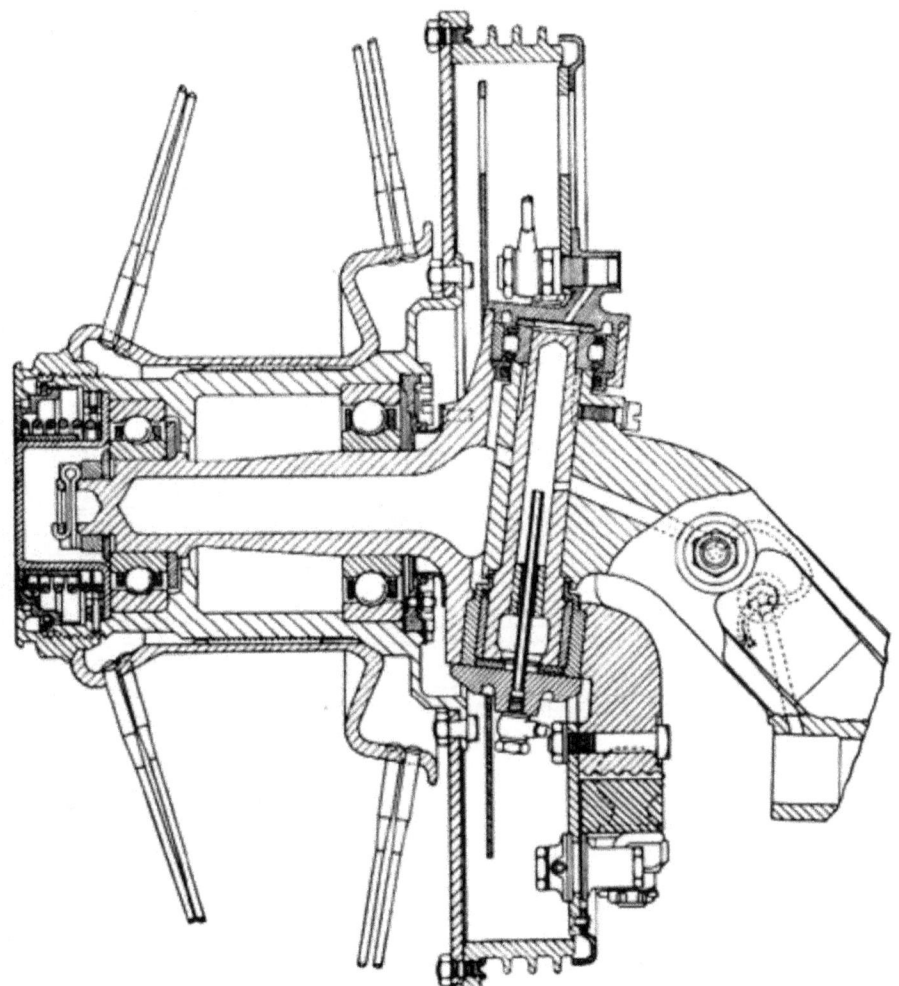

Fig. 16. - SECTION OF FRONT HUB

The thread of the hub nut is right-handed for all wheels

When replacing a wheel, care must be taken that the engaging surfaces, serrations and threads of both hub and wheel are free from road grit and other foreign matter. Preferably, they should be slightly greased.

The hub nuts must be tightly screwed up by means of the special spanner, and the use of the mallet in conjunction with it, to ensure absolute tightness.

The locking plate should now be allowed to come forward by turning the small lever, **P**, in an anti-clockwise direction, in order that its serrations shall engage those of the hub nut.

It should be observed that when jacking up a rear wheel, care is necessary that the head of the jack is arranged in the proper position. It should be imediately beneath the axle, between the two "U" bolts which secure the axle and spring together.

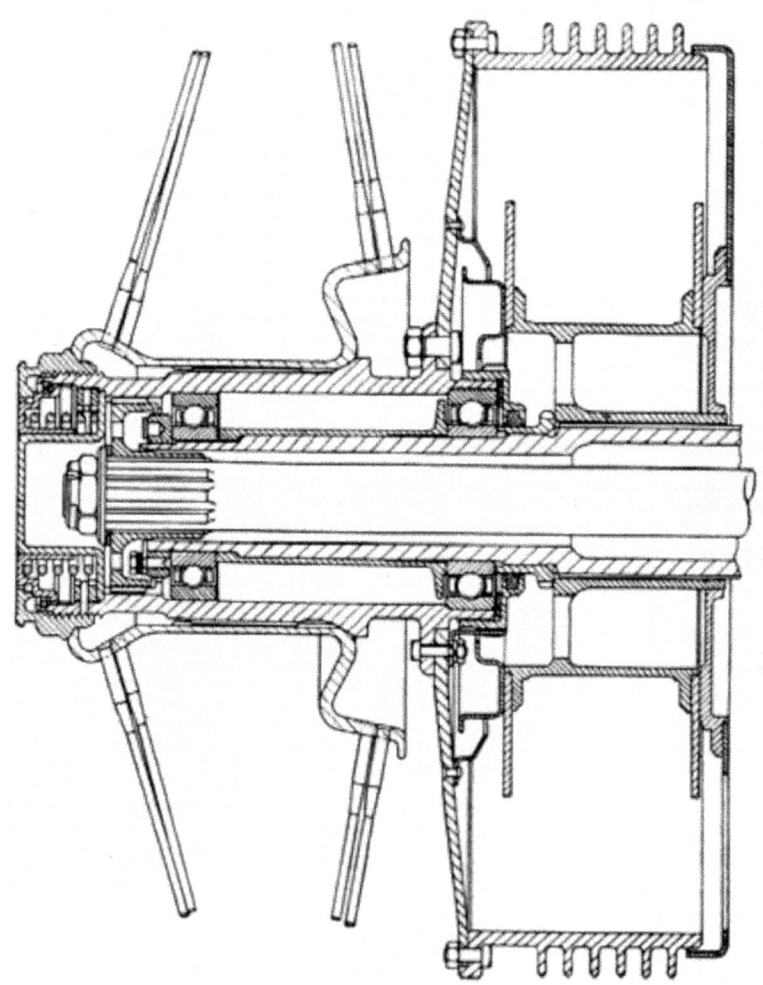

Fig. 17 - SECTION OF THE REAR HUB.

Care of Wheels.

Every 2,000 miles, the hub nuts should be tested for tightness with the spanner.

PERIODIC LUBRICATION AND ATTENTION.

On no account should the car ever be run with a wheel even slightly loose, as this will cause irreperable damage to the serrations and screw threads.

It is necessary to try each hub nut periodically with the spanner, and tighten if necessary. In order to tighten the hub nut, it is necessary for the locking plate to be forced back by means of rotation of the small lever **P** until its serrarions are disengaged from those of the hub nut.

Care must be taken when driving close to a high curb to avoid catching the projecting spokes of wire wheels. Very serious damage may thus be done to the wheel.

Lubrication of Wheel Bearings.

The wheel bearings are filled with ball-bearing grease in the first instance, and should run a long period without attention.

Sections of the front and rear hubs are given in Figs. **16** and **17** respectively.

Replacement Tyres.

Either Dunlop "Fort" or India "Super", size 6.00" x 19" tyres are suitable.

When ordering new tyres, the above should be specified. With regard to the inner tubes, it is necessary to state the size and to mention "well-base".

Tubes made for flat-base rims should not be used.

Balancing the Road Wheels.

It is most important, in view of the high speeds attainable, that the front road wheels should be properly balanced. Therefore it is necessary to have all wheels balanced and to re-balance a wheel after changing its tyre.

An out-of-balance effect is usually present in the complete wheel and tyre due to:-

(a) The valve and its patch on the inner tube;
(b) the joint of the inner tube; and
(c) unavoidable irregularities in the outer cover due to movement of the material during vulcanising.

A red spot on the outer cover wall indicates its lightest part, and the cover should be fitted so that the red spot is at the valve position.

To correct such out-of-balance, three bolts are provided, spaced at equal intevals around the wheel rim, as shown at **S**, **T** and **U** in Fig. **18**, and each carries a number of lead washers, enclosed by a metal cover.

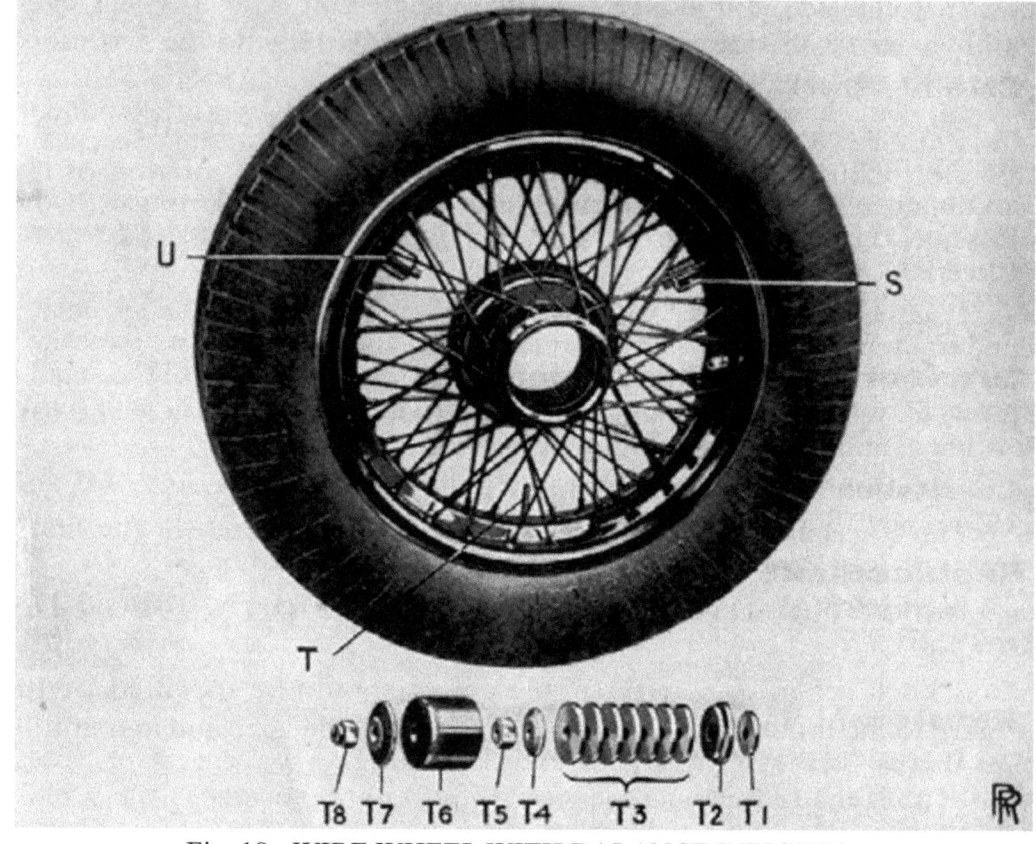

Fig. 18 - WIRE WHEEL WITH BALANCE WEIGHTS.

One of the bolts **T** is shown with its cover and washers dismantled. The parts are assembled on the bolt in the following order:-

1. Rubber washer **T**1, which acts as a seal against the ingress of water.
2. Special steel washer **T**2 which forms a firm base for the cover and the lead washers.
3. Lead balancing washers **T**3 up to seven in number on any one bolt.
4. Steel washer **T**4.
5. Nut, **T**5 for retaining lead washers.
6. Cover, **T**6.
7. Steel Washer, **T**7.
8. Cap nut, **T**8 for retaining cover.

To balance a wheel, all the lead washers should first be removed from each bolt, the other parts being fitted as indicated above.

The front axle being jacked up, the wheel must be turned gently and allowed to come to rest.

The lowest point of the tyre should then be marked.

The operation should be repeated, and if the original mark returns to the bottom position, one or more lead washers should be added to the bolt on the opposite side of the wheel.

If the mark made on the tyre is adjacent to the bolt, then one lead washer should be fitted on each of the other two bolts.

On the other hand, if no bolt should lie on the vertical centre line through the marked point on the tyre, the washers of the two bolts farthest from the mark must be altered, for instance, if the distance of one bolt from the centre line is approximately twice that of the other, two lead washers should be fitted on the bolt nearer to the centre line and one lead washer on the other bolt.

This process should be continued until the wheel will remain in any position in which it may be brought to rest, the number of lead washers being kept down to a minimum consistent with good balance of the wheel.

CHAPTER III
Fuel System

Fuel Feed - Fuel Filters - The Carburetter (previous to Chassis GYD-25) - Cleaning of the Air Valve-Setting of the Jets - Mixture Control- Slow Running - Starting Carburetter - Float Feed Mechanism - Crankcase Btreather Pipe to Carburetter - Dismantling the Carburetter - The Carburetter (Chassis GYD-25 and onwards) - Faulty Adjustment of Carburetter - Idling and Low-Speed Adjustment - Automatic Air Valve - Float Feed Mechanism.

Fuel Feed.

The fuel feed is arranged on the system by which the vacuum induced in the induction pipe of the engine raises the fuel from the main tank situated on the back of the car to a small service tank on the engine side of the dash, when it flows by gravity to the carburetter float chamber.

A cork-seated change-over tap is located on the dashboard and controlled from the driver's side, having its dial plate marked **R** (Reserve), **O** (Off), and **M** (Main). In the off position the supply from the main tank to the vacuum tank is shut off, and also that from the vacuum tank to the carburetter.

Some chassis have a 14 gallon main tank with a reserve of 2 gallons, while other chassis have an 18-gallon main tank with a reserve of 2½ gallons.

A catch must be depressed before the tap can be turned to the **R** position.

If the main fuel supplu be exhausted during a run, it should be observed that the service tank will also have been emptied, and after filling the main tank, the service tank must also be recharged before the engine can be started. This can be done by cranking over the engine for a few revolutions, a depression thereby being induced in the induction pipe, which will draw up fuel from the main tank into the service tank.

Fuel Filters.

On the earlier models, Chassis Nos. GXO11 to GYD-23, a small conical filter gauze is located on top of the inlet chamber at the junction of the main petrol supply pipe to the vacuum supply tank, and irregularity in the wotking of the vacuum feed may be due to choking of theis filter with foreign matter.

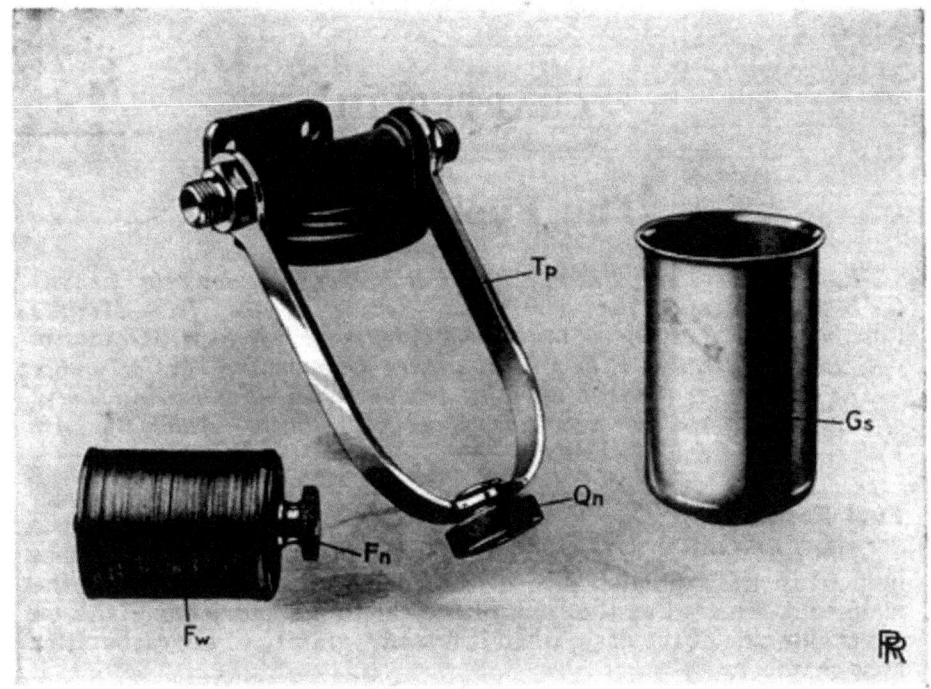

Fig.19. DASHBOARD FILTER DISMANTLED.

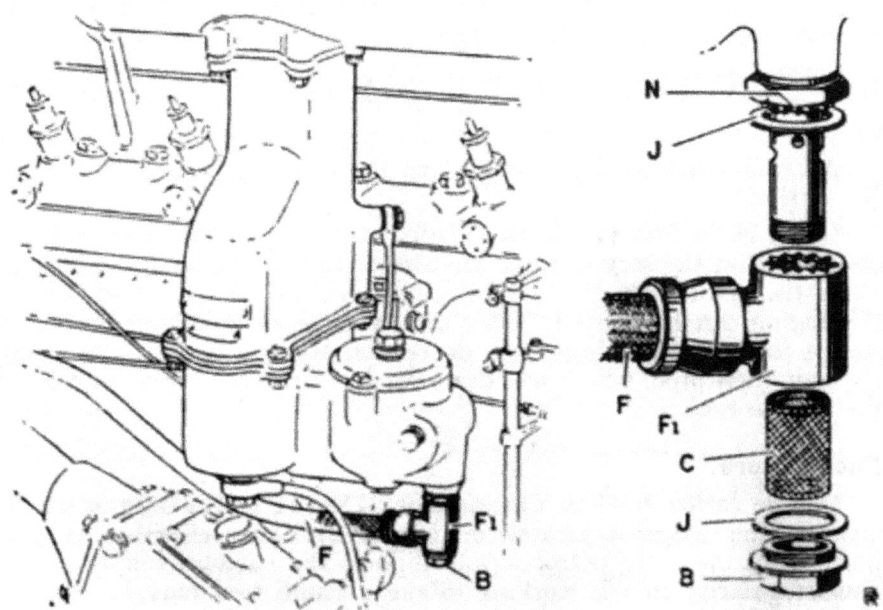

Fig.20 - CARBURETTER FILTER.

FUEL SYSTEM

It should be removed and carefully cleaned every 5,000 miles as directed on page 28.

A filter is arranged on all models, in the pipe line between the vacuum feed tank and the carburetter float chamber, and is mounted on the front of the dashboard, under the steering column.

The filter is shown dismantled for cleaning in Fig. **19**. It should be cleaned every 5,000 miles, as directed on page 28.

On the later models, Chassis GYD-25 and onwards, a small gauze strainer is arranged in the fuel inlet to the float chamber of the carburetter.

This should be removed and cleaned every 5,000 miles as directed on page 28.

On certain chassis two strainers are arranged in the main fuel tank, as shown in Fig. **21**.

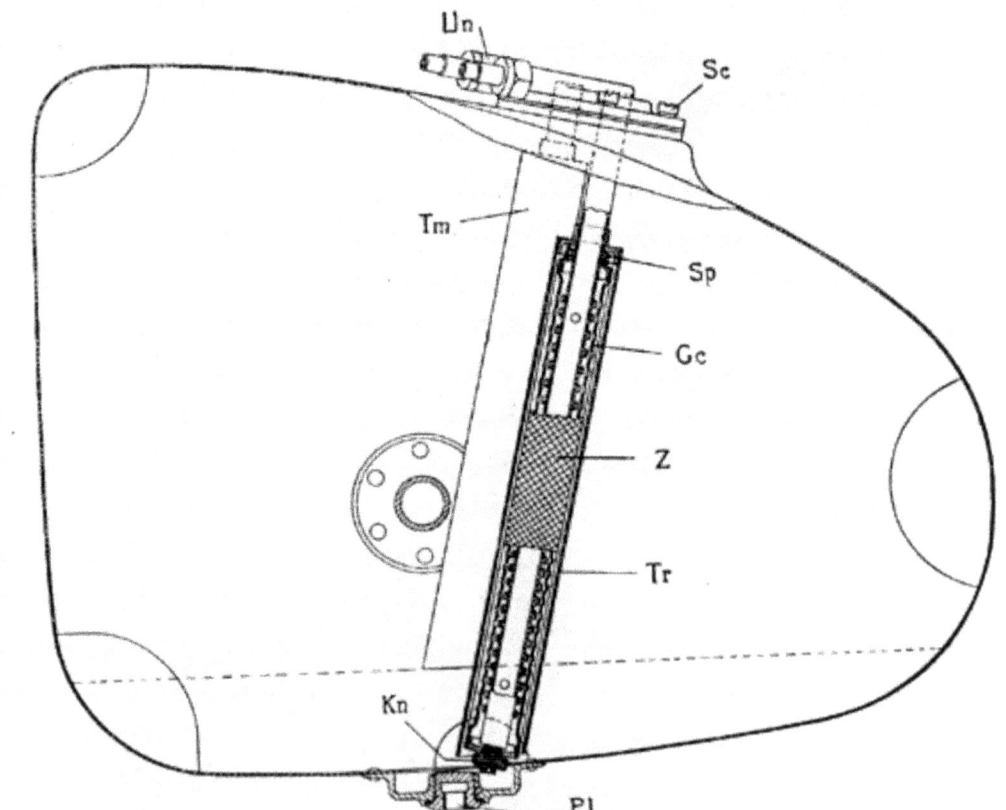

Fig. 21. - FUEL TANK STRAINERS

To remove them for cleaning every 5,000 miles as directed on page 28, the unions **Un** should be disconnected and the six screws Sc, removed.

The strainers may then be lifted out of the tank, care being taken not to damage the leather joint washer.

Each gauze is retained in position by a knurled nut **K**n, which is prevented from coming adrift by a split cotter. After removal of the latter, and unscrewing the knurled nut, the gauze may be removed, care being taken not to lose the small coil spring, which is arranged on top of the gauze carrier.

Remove and clean the gauze by washing in petrol or parrafin, using a brush.

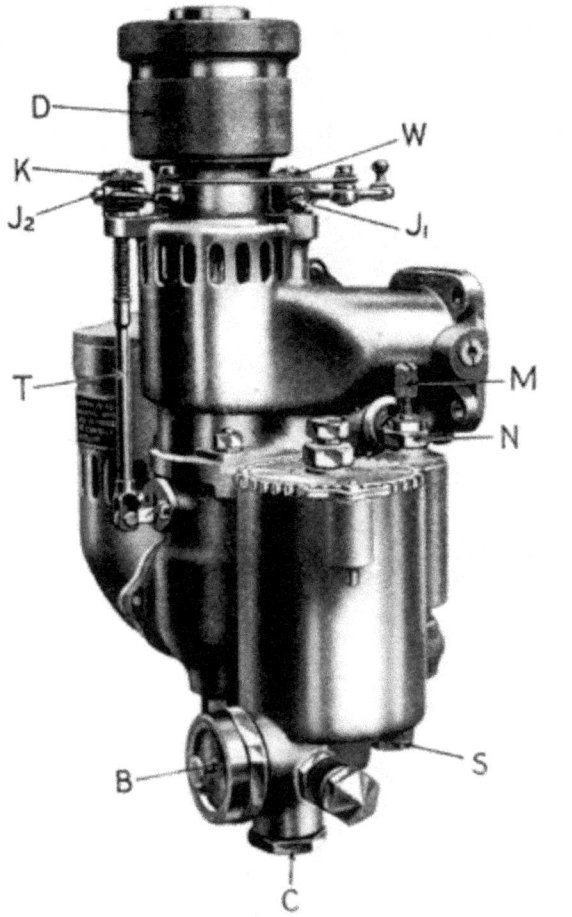

Fig. 22. - CARBURETTER (Chassis previous to GYD-25.)

B. Filter.
C. Drain Plug.
D. Piston Cylinder.
J1. Clamping Screw.
J2. Clamping Screw.
K. Knurled Nut.
M. Regulator - Starting Carburetter Jet.
N. Piston - Starting Carburetter.
T. High-speed Jet Control.
W. Low-speed Jet Control.

FUEL SYSTEM

When rplacing the strainers in their tubes, care must be taken that the coil springs are in position. The knurled nuts are intended for turning with the fingers only, and on no account must any tool be used on them. After they are replaced, brass split cotter must be fitted in the holes provided.

When replacing the strainers in the tank, care must be taken that the leather joint washer is in position.

The Carburetter. *(For Chassis previous to GYD-25.)*

The carburetter is of the Rolls-Royce automatic expanding type provided with two jets adjustable by a single lever under the driver's control.

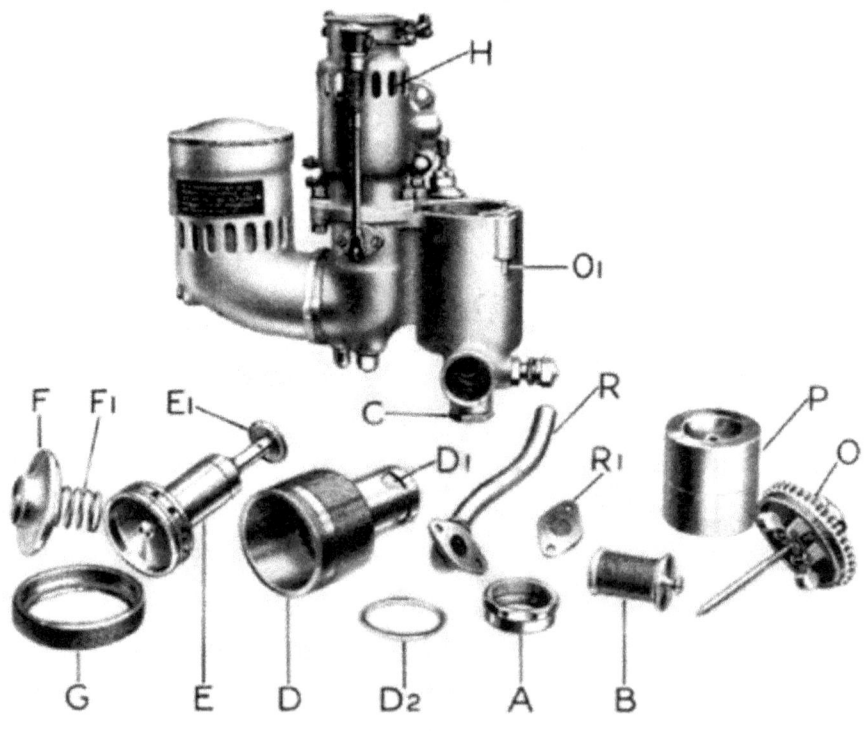

Fig. 23. - CARBURETTER WITH CERTAIN PARTS DISMANTLED.

- A. Retaining Nut - Filter.
- B. Filter.
- C. Drain Plug.
- D. Cylinder.
- D1. Cylinder Ports.
- D2. Joint Washer.
- E. Piston.
- E1. Diaphram.
- F. Cap - Cylinder
- F1. Spring.
- G. Retaining Ring - Cap
- H. Air Ports.
- O. Cover - Float Chamber
- O1. Catch - Cover.
- P. Float
- R. Crankcase Breather.
- R1. Gauze Filter.

Each of these jets is located in a venturi tube, the smaller one always being in action, and the larger one being automatically brought into action by an increase, beyond a certain value, in the depression existing within the carburetter, due to an increase of engine speed or throttle opening, or both.

The complete carburetter is shown in Fig. **22**, and in Fig. **23** it is shown with certain parts removed.

The outlets of the jets are regulated by taper needle valves, that for the small or low-speed jet being shown at **W**1 (Fig. **22**), and the control for the large or high-speed jet needle at **T**.

The automatic expanding effect is attained by the provision of a suction-operated piston working in a cylinder, **D** (Figs. **22** and **23**), located above the high-speed jet.

The cylinder **D** and piston **E** are shown removed for cleaning in Fig. **23**. The cap **F** carrying the spring **F**1 fits over the top of the cylinder, and is retained by a knurled nut **G**.

Increased depression in the carburetter raises the piston **E** against the spring **F**1, carrying with it a diaphram **E**1, which fits into, and in its lowest position, blanks off the larger choke tube. The lifting of this diaphram admits air past the high-speed jet.

More movement of the piston not only opens the high-speed choke tube still further, but also admits air by uncovering the ports **D**1, and air gaining admission through ports **H** in the carburetter, thereby counteracting the tendency for the mixture to become over-rich at increased air velocity.

The various adjustments should on no acount be altered, the carburetter having been carefully set by the makers in the first instance.

The mixture control lever, which operates on both jets simultaneously, provides ample range to suit ordinary variations in running conditions, such as different atmospheric temperatures and different fuels, including the use of benzole or benzole-petrol mixtures.

Cleaning the Air Valve.

The air valve valve and cylinder should be removed every 2,500 miles and carefully wiped with a piece of dry cloth, as directed on page 21.

No oil should be used on the valve or its cylinder.

It is advisable when replacing these parts to refit the cylinder to the carburetter without the air valve, the latter being replaced afterwards.

Care should be taken when replacing the cylinder **D** to see that the metal washer **D**2 is in position and its joint faces are perfectly clean.

It must be emphasised that great care is necessary when handling these parts, as they have been machined to fit very accurately, and any slight distortion is liable to impair the working of the carburetter.

Setting of the Jets.

If the adjustment of the jet needles has been upset for any reason it acn be restored in the following manner: -

FUEL SYSTEM

With the mixture control lever set half-way along its quadrant and the clamping screws **J**1 and **J**2 (Fig. 22) of the jet needle levers slack, each of the knurled nuts **K** and **W** should be turned until the line filed across them registers with the line across the end of the corresponding screwed spindle, the end of the spindle being at the same time flush with the end of the nut.

The clamping screws **J**1 and **J**2 should then be tightened, and the makers' setting will have been restored.

If, owing to damaged and replacement parts, it becomes necessary to re-set the jets with no guide in the form of markings referred to, it is strongly recommended that the makers should be consulted, and this work not attempted without their advice or assistance.

In the eveny,however, of circumstances rendering such a course impossible, or very inconvenient, proceed as follows: -

With the mixture control lever set half-way along its quadrant and the clamping screws **J**1 slack, the knurled nut **W** should be turned in a clockwise direction until its lower side just commences to lift away from the facing against which it normally rests.

The low-speed jet will now be fully closed, the tapered part of the needle resting on the mouth of the standpipe.

A preliminary setting can then be obtained by rotating the nut **W** in an anti-clockwise direction through approximately one complete turn. The clamping screw **J**1 should then be tightened.

In the case of the high-speed jet it is noy practicable to obtain a preliminary setting in this way because the tapered portion of the high-speed jet needle is arranged to pass freely inside the bore of its standpipe. This is done in order to protect these parts from damage which otherwise might result if the nut **K** were turned to force the taper of the jet needle into the standpipe.

Consequently no visible indication is available to show precisely when the high-speed jet is fully closed, and it will be necessary to discover its approximate position by running the engine.

It will be possible to start up the engine after setting the low-speed jet needle as described, and this should now be done, the mixture control lever being set half-way along its quadrant.

If, when the throttle is opened moderately by means of the lever on the steering wheel, the engine pops back through the carburetter and possibly stops, the mixture is too weak, and if black smoke comes from the exhaust and the engine misses fire and perhaps stops, the mixture is too rich.

To weaken the high-speed jet setting, the screw **J**2 should be released and the nut **K** turned in a clockwise direction; and to strengthen it nut **K** must be turned in an anti-clockwise direction.

Having arrived at a preliminary setting for the high-speed jet in this way, and with the mixture control lever again set half-way along its quadrant, the throttle should be opened by means of the lever on the steering wheel until a speed is reached at which the automatic piston valve is on the point of lifting but has not actually lifted. Movement of this can be observed by looking through the air ports in the carburetter.

The clamping screw **J**1 of the low-speed jet needle should then be slackened, and the knurled nut turned in a clockwise direction until the engine speed becomes slightly reduced.

The clamping screw should then be tightened, and the mixture control lever moved first over to strong and then to weak. If in *both* of these positions the engine hesitates, or even possibly stops in the weak position, then the adjustment of this jet is fairly correct.

To test the high-speed jet setting the acelerator pedal should be depressed momentarily, and the lever again tried in both its extreme positions. In either position a distinct loss of power should be experienced. If these variations do not occur, or occur in only one of the extreme positions of mixture strength, the settings should be varied accordingly by slackening the clamping screw and turning the high-speed knurled nut in a clockwise or anti-clockwise direction, according as the mixture requires weakening or strengthening respectively.

The foregoing will only provide an approximate or trial setting.

When the car is taken on the road for final adjustment, the driver should bear in mind that the high-speed jet comes into operation at about four miles per hour in top gear on the level. Consequently any sign of too rich or too weak a mixture below this speed is an indication that the low-speed jet requires adjustment.

At speeds above four miles per hour, the high-speed jet has an increasing influence over the mixture.

The best all-round setting of the jets is one in which movement of the mixture control lever to either of its extreme positons will, at any speed, cause a distinct loss of power and possibly misfiring. Steady running and good power at all speeds should be obtained with the lever set half-way on its quadrant.

Mixture Control

Utilised in a proper manner, very economical running can be obtained.

When starting the engine from cold, especially in cold weather, the mixture lever should be moved over to **Strong** before changing from the starting to the main carburetter.

As the engine warms up it will be found that the lever can be moved towards the half-way position, until, with a well-warmed engine and normal touring conditions, it can be taken up a few notches towards **Weak.**

A weak mixture burns more slowly than a normal one, and to get the best power from such a mixture, the ignition needs to be well advanced. Consequently, the most economical running is obtained when the ignition lever is fully advanced and the mixture control set as

FUEL SYSTEM

far towards **Weak** as the conditions allow without seriously reducing the power available.

If, on the other hand, weakening of the mixture is carried too far, then, apart from the probability of misfiring and popping in the carburetter, similar road conditions will call for a bigger throttle opening, and the economy desired be thereby nullified.

Under severe conditions, such as a long ascent which calls for full throttle, too weak a mixture may cause overheating. So the control lever may with advantage be set a little **Strong** under these circumstances.

Slow Running.

The best slow running will be obtained with the mixture control set two or three notches **Strong.** If difficulty is experienced in getting the enfgine to run slowly, this may be due to the flow of petrol past the low-speed jet needle being restricted by the presence of foreign matter.

To remove this, the jet needle should be raised with the fingers by lifting knurled nut **W** (Fig. **22**), and the throttle simultaneously opened to race the engine momentarily.

If this effects a cure, it would be advisable to clean the petrol filters, as these are probably dirty.

The trouble may also be due to sticking of the carburetter air valve, or faulty tappet adjusment.

Starting Carburetter.

A special auxiliary jet and expanding choke tube is incorporated in the carburetter for starting purposes only.

This jet can be regulated by means of the knurled screw **M** (Fig. **22**), which carries a taper needle running into the jet. Turning this screw in a clockwise direction reduces the jet opening, and in an anti-clockwise direction increases it.

Should occasion arise to re-set this jet adjustment, the screw should be turned with the fingers in a clockwise direction until it is felt that the needle is entirely closing the jet. It should then be rotated in the opposite direction for about one-and-a-half complete turns. This will give a setting at which the engine can be started. Then, with the engine running, the screw may be turned to weaken or strengthen the mixture slightly as may be required.

It is important that the setting of the needle should not be such as to provide an over-rich mixture. Although an average setting is one-and-a-half turns from the closed position as stated, this may be reduced to one-and-a-quarter turns in warm weather. On the other hand, in very cold weather, it may be increased to one-and-three-quarter turns, but must be again reduced when the weather becomes warm.

Adjustment of the starting carburetter should only be performed when the engine is cold.

The variable choke or throat of theis small carburetter consists of a suction-operated piston, which is lifted against gravity and automatically adjusts the choke area to suit the engine speed.

Access to this throat is obtained by unscrewing the cap **N**, which may then be lifted out with the jet needle. It is advisable occasionally to remove and carefully wipe the piston, but no oil should be used on it.

As the succesful working of this small carburetter is dependent on an air-tight induction system, it is essential that the main throttle should be fully closed when starting the engine.

When changing over to the main carbutetter, the throttle should be moderately opened and the starting carburetter lever turned to the **Running** or **Off** position, where it should always remain, except for starting. If the engine hesitates or tends to stop, the starting carburetter should be opened again and the main throttle closed until the temperature conditions of the engine are suitable for steady running on the main carburetter.

Cases have arisen of piston seizure which have been traced to excessive use of the starting carburetter. It should be appreciated that the object of the starting carburetter is to facilitate starting when the engine is quite cold, the mixture it provides under such conditions being on the rich side. Consequently, excessive use of the starting carburetter, or its use with a hot engine, is liable to cause liquid petrol to be drawn into the cylinders and wash away the engine oil.

Further, if used with a hot engine, starting may be difficult, due to the over-rich mixture.

The starting carburetter should not be used for more than half a minute before changing over to the main carburetter, and not used at all with a hot engine, in which circumstances starting will be found quite easy on the main carburetter only.

Float Feed Mechanism.

The float chamber should be cleaned out every 5,000 miles as directed on page 25, by unscrewing the cover **O** (after raising the catch **O**1, if such is fitted) and removing the float **P** (Fig. **23**). The interior of the float chamber should be wiped out with a piece of clean, damp wash-leather.

No provision is made for flooding the carburetter by agitating the float needle, as this is never necessary. The starting carburetter is provided to supply suitably rich mixture for starting purposes.

If flooding occurs, it is probably due to foreign matter having lodged on the needle valve seating, and steps should be taken accordingly.

FUEL SYSTEM

Crankcase Breather Pipe to Carburetter.

In order to reduce the emission of oil fumes from the engine, a pipe is carried from the crankcase to the carburetter air inlet.

This pipe is shown removed at **R** in Fig.**23**.

A small gauze, **R**1, is arranged between the pipe flange and the carburetter, which in course of time may require cleaning. It should be removed and cleaned every 10,000 miles, as directed on page 29.

Dismantling the Carburetter.

Normally it should not be necessary to dismantle the carburetter to a further extent than that already mentioned. On the other hand, it sometimes occurs that the jet needles become sticky in operation, due to sediment and impurities in the fuel, and the correct functioning of the carburetter is impaired.

Under such circumstances the carburetter should be removed bodily from the engine for dismantling.

The plugs below bioth jet needles should then be removed and cleaned of sediment. At the same time it should be ascertained that the spring plunger below the high-speed jet needle is working quite freely. The upward pressute of this spring is relied upon to open the high-speed jet, and its freedom of movement is therefore of great importance.

After removing the air valve and its chamber two countersunk set-screws near the low-speed jet needle should be unscrewed. The jet needle can then be carefully lifted out.

The high-speed jet needle is removed by taking outthe pin from the jaw at the lower end of control rod **T**(Fig. **22**) and unscrewing the two countersunk set-screws which secure the bearing of the operating lever to the side of the carburetter. The needle jet may then be lifted out.

It is advisable to clean both jet needles carefully in paraffin. The jets themselves should also be cleaned out by using a small wooden stick and a piece of rag soaked in paraffin.

There should be no need to separate the two parts of the carburetter body, but if this is done, it is of vital importance to remove the air valve and its chamber first of all, and also the low-speed jet needle. The latter will almost certainly be damaged if left in position when the carburetter body is divided.

The Carburetter (For Chassis GYD-25 and onwards.)

The carburetter is of the single jet expanding type, and is of Rolls-Royce design and manufacture.

Reliable idling of the engine is ensured by the provision of "Throttle-edge" carbutation.

Air passes from the air cleaner and silencer into the carburetter in a horizontal direction across the main jet orifice, the latter being regulated by a taper needle valve attached to the air valve. A trunk or extension carried by the air valve protrudes into the air-way above the jet and acts to vary the effective choke area.

The position of the air valve when the engine is running is determined by the depression of the induction pipe, the valve being connected by passage ways to a point between the butterfly throttle valve and the carburetter.

The air valve assembly is normally prevented from falling lower than a certain point by means of a stop which encircles the jet needle and abuts the main jet orifice. When the engine is idling or operating at small throttle openings, the air valve rests on this stop. An independent, miniature carburetter having a fixed jet and adjustable choke then comes into action, its mixture being delivered through

Fig. 24 - CARBURETTER IN POSITION ON ENGINE

FUEL SYSTEM

a small hole adjacent to the edge of the butterfly throttle when this is nearly closed. The position of this small discharge hole relative to the edge of the throttle is arranged to be adjustable, in a manner to be described, in order to secure reliable idling and a smooth "change-over" from the idling jet to the main jet on opening up.

To facilitate starting from cold, there is a control lever on the instrument board, its quadrant being marked **Start** and **Normal** which, in the **Start** position, operates to lower the main jet orifice and thereby to permit the air valve to fall lower. This causes the choke area to be reduced, so providing a rich mixture for starting. As the jet orifice is increased relative to the choke area, the enriching effect is maintained, though to a decreasing extent, as the throttle is opened and the air valve lifts. The control, however, is only intended for use when starting from cold.

In Fig. **24**, the carburetter is shown in position on the engine, and below it is shown detached from the engine and with certain parts dismantled.

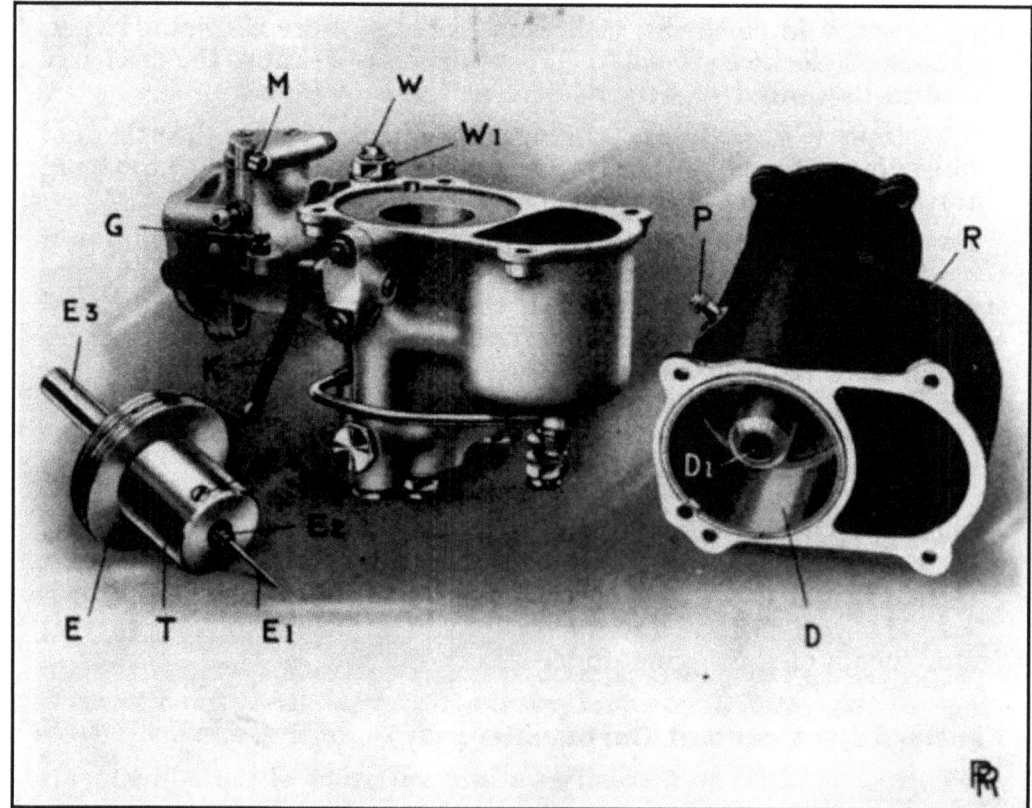

Fig. 25.- CARBURETTER, WITH CERTAIN PARTS DISMANTLED.

Formed integrally with the air inlet **R** is a cylinder **D** in which operates the air valve **E**. this carries the taper needle valve **E**1 for control of the main jet orifice, and a trunk or extension **T** to adjust the air passage or choke in unison. A perforated sleeve **E**2 acts as the stop referred to and limits downward movement of the air valve under ordinary running conditions. There is a clearance between the air valve and its cylinder so that these parts do not actually touch, in order to avoid sluggishness in operation due to friction, the location of the piston being effected by an integral central extension **E**3, which closely fits a guide **D**1 in the cylinder.

The control lever for the main jet orifice is shown at **K**. As explained, this is only used for starting from cold, being controlled by a thumb lever on the instrument board.

The adjustable choke of the idling and low-speed carburetter is shown at **W**, the adjustment being locked by the lock nut **W**1.

The point of discharge of the idling mixture into the main airway relative to the throttle valve is rendered adjustable by forming the orifice eccentrically in a loose plug. The latter can be rotated, and the position of its hole thereby varied, by turning the screw **M** with a screwdriver. The influence of this adjustment is only felt when the engine is idling at a very low speed. Under these circumstances turning screw **M** clockwise will result in the mixture becoming richer, and counter-clockwise weaker. These movements cause the discharge orifice to be shifted slightly relative to the throttle edge.

A stop-screw **G** limits closing movement of the throttle and enables this to be set so that the engine will idle reliably with the hand throttle lever at the bottom of its quadrant.

As the throttle is opened from the idling position the location of the adjustable discharge orifice has a decreasing effect on the mixture strength, but on the other hand, the adjustment of the low-speed choke has a greater effect. The influence of this in turn becomes progressively reduced as the throttle is opened still further and the main jet comes into operation. This occurs when the engine is idling at about 750 r.p.m. Further throttle opening or increase of engine speed causes the air valve to lift, thereby increasing the fuel supply by lifting the needle **E**1, and the air supply by raising the trunk **T**. Similarly, it falls as the throttle is closed or the engine speed falls, a state of balance being maintained whereby the air valve keeps at a certain height dependent on thwe engine speed and throttle opening. Thus, the carburetter automatically adapts itself to the varying requirements of the engine under all conditions.

Faulty Adjustment of Carburetter.

There should be no necessity for any variation of the adjustments of the carburetter as fixed by the makers. Great care is taken during testing of the car to secure the best settings, and these should not, under normal circumstances, be altered.

It is realised, however, that information as to the methods for restoring adjustments may prove valuable under special circumstances, such as

FUEL SYSTEM

accidental derangement or damage, and is consequently given, as far as is practicable, in the following paragraph.

Idling and Low-Speed Adjustment.

There are only three external adjustments, namely, the throttle stop screw **G**, the screw **M** for adjusting the position of the idling orifice, and the low-speed choke adjustment **W**. All of these are concerned with idling and low-speed running conditions.

Assuming these adjustments have been disturbed for any reason, the following procedure should be adopted when restoring them:-

The lock nuts **W**1 should be released, the low-speed choke **W** screwed in a closckwise direction as far as possible (in which position it will be resting on the low-speed jet) and then unscrewed one complete turn and be locked in this position by means of nut **W**1. The throttle stop screw **G** should be turned with a screwdriver, after releasing the lock nut, until it is clear of the lever on the throttle spindle. It should then be screwed in until it just comes into contact with the throttle lever and turned one complete turn further.

A trial adjustment of the plug which carries the idling orifice can be obtained by setting this in the mid-position of its range of movement. The total movement of screw **M** is about fourteen complete turns, therefore mid-position is obtained by rotating it as far as possible in one direction and then turning it in the opposite direction about seven complete turns.

The mechanical settings described, though preliminary and approximate, will at least enable the engine to be started up, and this should next be done.

The throttle should then be opened by means of the hand control until the engine is running light at about 750 r.p.m. The low-speed choke **W** should next be screwed in or out until the engine is observed to be firing evenly. The throttle should now be closed until the engine is running at a reasonably low idling speed, and screw **M** rotated one way or the other until the engine is firing perfectly regularly, as may be judged by listening to the exhaust pulsations.

Finally, the throttle stop screw **G** should be adjusted until the engine runs at a reliable idling speed in the neighourhood of 250 r.p.m.

The three adjustments described are best set finally when the engine is warm. Further, it will be found that the adjustment of the low-speed choke **W**, and that of the idling orifice **M**, are to some extent inter-dependent. Any sign of hesitation on opening up on the road at low speeds is probably due to weakness of the low-speed carburetter. This can be corrected by turning the screw **W** clockwise a little, thus reducing the low-speed choke area.

Automatic Air Valve.

The air valve requires no attention beyond lubrication of its guide.

If defective running should develop, as evidenced by serious hesitation on pick up and posibly popping in the carburetter, it is probable that the air valve is sticking slightly. In such circumstances the air silencer H must be taken off, but before doing this the breather pipe between the silencer and the valve rocker cover must br removed with the fingers. The silencer can then be taken off after removing nuts H1 and H2. Next the bolts R1 should be removed preparatory to lifting off the inlet pipe R. Great vare must be taken when doing this that the air valve does not drop out and become damaged. The best plan is to lift off the pipe R with both hands and insert the fingers at either side beneath the lower face of the pipe as soon as this is raised sufficiently. The valve can then be prevented from falling out by the fingers of either hand, the pipe being raised vertically until the trunk and needle of the air valve are clear of the carburetter body.

The valve E, its extension E3, and the guide D1, should be carefully wiped with a piece of clean cloth dipped in petrol and the guide lubricated woth a few drops of thin oil.

No oil should be used on the piston valve or its cylinder, and no polishing paste or abrasives used to clean these parts. the utmosr care must be taken not to bend or damage the depending needle valve E1 or to bruise the valve in any way.

When replacing the air valve it will be noticed that there is a slot in the trunk T, which must engage a small projection on the carburetter body.

A lubricator P is provided for lubricating the guide. Every 5,000 miles, as directed on page 25, the cap of this lubricator should be turned until the oil hole is exposed and one or two drops of clean, thin oil injected. Care must be taken afterwards to close the lubricator in order to exclude dirt ot grit.

The needle valve is secured in position by means of a grub screw E4. If it should be necessary to remove this - as for instance, when replacing an accidentally damaged needle - care must be taken that the needle is gently pushed into the trunk of the air valve as far as possible and the grub screw E4 tightened.

If a needle should be accidentally damaged, a new one must be obtained from Rolls-Royce Ltd. A number is stamped on the end of the needle. On no account must one of another size be fitted.

Float Feed Mechanism.

The float feed mechanism is of the pivoted ball type. The cover is secured by two screws, a paper joint washer being fitted between the cover and the carburetter body.

No attention should be necessary to these parts.

A flexible pipe F conveys fuel from the dashboard filter to the float chamber. *On no account must the fitting on the pipe itself be disturbed.*

CHAPTER IV

The Brakes

General - Adjustment of Rear and Front Brakes (Chassis previous to GAF-1) - Adjustment of Foot and Hand Brakes (Chassis GAF-1 and onwards) - Adjustment of Servo

General.

The only points in the sytem where any adjustment is provided or is necessary are the following: -

(i) **Rear Brakes** *(Chassis previous to GAF-1).*
The threaded rods coupled to the cam operating levers below the ends of the rear axle

Rear Brakes *(Chassis GAF-1 onwards).*
A wing nut **Wr**, for adjustment of the foot brakes, and the threaded rods coupled to the cam operating levers, Fig. **29**, for adjustment of the handbrake.

(ii) **Front Brakes** *(Chassis previous to GAF-1).*
A serrated adjustment on the cam operating shafts.

Front Brakes *(Chassis GAF-1 onwards).*
A wing nut adjustment, see Fig. **28**.

(iii) **Servo.**
A serrated adjusting nut on the end of the servo shaft.

It is very important to observe that under no circumstances should adjustment be attempted at any other points, for instance, by altering the lengths of some of the brake rods or of any of the ropes.

Any alterations to the lengths of these rods or ropes will virtually shorten the lengths of some of the levers, and will interfere with the correct functioning of the system.

Adjustment of Rear Brakes. *(Chassis previous to GAF-1).*

The state of adjustment of the rear brakes - both foot and hand-operated - should be tested by reference to movement of the brake cables necessary to take up the clearance between shoes and drums, or to the movement at the ends of the levers on the axle to which the cables are connected. For this purpose the cable should be pulled or the lever operated by hand and the movement measured. This movement should never be less than 1" for both foot and hand brakes, but there is no need to adjust the brakes unless it exceeds 1½" for the foot brake, or 1¾" for the hand brake.

The metohod of adjustment is similar for both hand-operated and rear foot-operated brakes, and is illustrated in Fig. **26**. This is a view looking from the underside of the axle.

Fig. 26. - REAR WHEEL BRAKE ADJUSTMENT.
(View from Below)

The outside rods **Y** actuate the foot brake shoes, and adjustment is effected by removing the pin **Y**1 from the jaw **Y**2, this pin being secured by a collar and split pin cotter, slackening the small nut **Y**3, and screwing the jaw farther on to the rod **Y** to an extent depending on the amount of adjustment required.

The amount of adjustment made to both these rods should usually be the same. A convenient method of checking this is to measure the distance between the collar **Y**4 and the jaw **Y**2.

Before replacing the pins **Y**1 in the jaws, attention should be turned to adjustment of the hand brake, if any is required.

All adjustment for the hand brake is made on the inside rod **H** and the corresponding rod at the other end of the axle.

THE BRAKES

The adjustment is effected in a similar manner to that of the foot brake, but it should noticed that the pin **H**1 of the hand brake jaw **H**2 cannot be removed until jaw **Y**2 is disconnected.

Care should be taken that the collar which fits the pin of each jaw is in position before fitting the split cotter.

The adjustment of both brakes should be checked finally by measuring the travel of the cable, as already described, when the cable or lever is moved from the off position to a point where the shoes just touch the drums.

After replacing the pins and their collars, split cotters should be fitted to these, and the samll nuts **H**3 and **Y**3 tightened up.

The amount of adjustment provided is so proportioned that when all has been utilised (jaws **H**2 and **Y**2 being against the collars **H**4 and **Y**4 respectively), it is a sign that the brake shoes require recovering, and the makers or one of their "Special Retailers" should be consulted.

On no account should further adjustment be attempted as, for instance, by shortening the brake ropes or interfering with adjustments within the brake drums. Such a course might result in serious injury to the drums and shoes.

Adjustment of Front Brakes. *(Chassis previous to GAF-1)*.

It should be borne in mind that pedal travel is no indication as to the front brake adjustment, because these are entirely servo operated and their adjustment will not influence the pedal travel.

The only indication that they require adjustment (apart from an observed decrease in front braking) is excessive movement at the end of the levers **L**f, Fig. **27**. When lightly depressed by hand the movement at the end of this lever, for correct adjustment, should be about 9/16". It should not exceed ⅞".

When this figure is exceeded adjustment is imperative.

It is effected as follows: -

Remove the split cotter of the castellated nut **J**1 and unscrew the latter. The cover **J** may then be removed, exposing the serrated adjustment. As this cover also acts as a locking piece, it will be found convenient to mark the position of engagement of its teeth with those on the member **R**f before removing it.

The nut **G**f should be unscrewed sufficiently to permit the serrated member **R**f to be moved clear of similar serrations on the lever **L**f. These two sets of teeth are marked respectively with an arrow and figures 0, 1, 2, 3, 4,and 5. If the brakes are being adjusted for the first time the arrow will point to "0".

Having noted the relative positions of these serrated parts, they may be disengaged by tapping the lever **L**f away from the wheel, carrying with it the serrated member **R**f. While holding the latter in the hand, the lever should then be tapped towards the wheel again, when the serrations will be disengaged.

Fig. 27. - FRONT WHEEL BRAKE ADJUSTMENT.

The cam operating shaft, and with its member **Rf**, should next be turned by means od a spanner on the hexagon **K** of the shaft until the parts can be re-engaged one serration further towards the on position of the cam operating shaft than before; that is, after the first adjustment the arrow will point to "1".

Finally, re-tighten the nut **Gf**, re-fit the cover **J**, which also acts as a locking piece for this nut, and replace the castellated nut **J**1, fitting a split cotter to the latter.

If any difficulty is experienced in getting the teeth of cover **J** to engage with those of member **Rf**, the cover should be rotated slightly and tried in different positions.

The brake clearances should be tested again after adjustment by measuring the movement of levers **L**f, as described. This movement musnot be less than 9/16", otherwise the brakes may drag.

THE BRAKES

Usually it will be necessary to adjust each front brake a like amount.

It should be observed that when the five teeth of adjustment have been utilised, this is an indication that the shoes require new facings.

On no account should further adjustment be attempted by, for instance, interfering with the lengths of any of the brake rods or ropes.

Apart from testing for the need for adjustment of the front brakes, it is important to test from time to time that the shafts and joints on the axle are free by pushing down the levers **L**f with the hand, or by moving levers **B** similarly.

The mechanism should feel free, and be returned sharply to the off position by the pull-off springs.

If any tightness is found, the cause must be investigated and removed, otherwise there is a danger of the brakes dragging and becoming damaged.

Adjustment of Foot Brakes. *(Chassis GAF-1 and onwards).*

The wing nut adjustment for the front brakes is shown at **W**f, Fig. **28**, and that for the rear foot-operated brakes at **W**r, Fig. **29**.

Fig. 28. - FRONT BRAKE ADJUSTMENT.

Only the fingers must be used in turning these nuts. They are formed with cam-shaped bosses bearing on cylindrical trunnions in such a way that rotation of the nut through 90° from the position shown causes the brake shoes to be moved towards the drum as the cams ride over the trunnion. This movement is carefully predetermined, and is equal to the normal clearance between shoes and drum when the shoes are in the off position. Screwing on the nut through a further 90, that is, a total of half, allows the shoes to return to an off position, which is half a turn of the adjustment nearer to the drum. The adjustment is self-locking.

When making or testing the adjustment, it is preferable that the wheel should be jacked up and rotated by hand. One is then able both to hear and to feel when the shoes make contact with the drum.

The nut should be screwed up until the cam action described prevents further rotation owing to the shoes being applied to the drums. The setting will then be correct, and the adjustment locked if the nut be turned back one-quarter of a turn.

It must again be emphasised that on no account must force be used in turning the nuts, as this will defeat the object of the described arrangement and result in badly adjusted, probably dragging brakes.

Movement of the brake pedal when the car is standing does not apply the front brakes, which are operated solely by the action of the servo. Under such circumstances, pressure on the pedal will only apply the rear brakes.

Adjustment of Hand Brakes. *(Chassis GAF-1 and onwards).*

All adjustment of the hand brake is effected at the outside rods beneath the rear axle, one of which is shown at **H**, Fig. **29**.

With the hand brake lever right off, the adjustment should be tested by pulling the brake rope **J** with the hand and measuring the travel of the rope necessary just to apply the brake. This travel should not be less than 1", but there is no need to adjust the brakes unless it exceeds 1¾".

Adjustment is effected by removing the pin **K** from the jaw, **L**, this pin being secured by a collar and split cotter, releasing the locknut, **H**I, slackening the small nut **M**, and screwing the jaw farther on to the rod **H** to an extent depending on the amount of adjustment required.

Usually, this should be the same at both sides.

Care must be taken to replace the pins **K**, securing them with split cotters and collars, then to re-tighten the nuts **M**, and finally the locknuts, **H**I.

Fig. 29 - REAR BRAKE ADJUSTMENT.

Adjustment of the Servo.

The servo is of the dry, disc-clutch type, and should run 20,000 miles without the need of any adjustment.

If adjustment is necessary, it is effected by screwing up the nut, **Z**, Fig. **30**.

This nut is locked by 25 rounded serrations formed on its face, which engage similar serrations on a washer, which is secured against rotation relative to the shaft. The depth of these serrations is carefully proportioned to give the correct clearance of the servo, the nut being turned so that the teeth lightly ride over each other and engage again.

On no account should force be used in this operation, as such treatment would nullify the object of the teeth, namely, ensure the correct clearance with very little trouble.

After effecting adjustment in this way, care should be taken to see that the serrations are in proper engagement.

The adjusting nut should not be screwed up more than one serration - that is, 1/25 of a turn - without testing the servo adjustment.

To test the servo adjustment the pedal should be depressed lightly by hand to engage the servo and compress the buffer springs, Z_1, but just short of moving the lever A_2 rotationally.

The pedal travel should then be not less than ½" measured at the top of the pedal towards the dash.

Fig. 30. - THE SERVO MOTOR AND ITS CONNECTIONS.

It must be realised that this movement is entirely due to operation of the servo, and does not alter the rear brake clearances. Hence, lever **A**2 is not moved rotationally, as mentioned.

After adjustment, the servo clearance should always be checked again by measuring the pedal movement, as explained.

Emphasis is laid on this point, as obviously a dragging servo, due to abuse of the adjustment provided, would result in dragging of the brakes on all wheels.

CHAPTER V

The Clutch

Clutch Pedal Adjustment.

There must always be a certain amount of "free" or idle movement of the pedal, it should be possible to lift it about ½".

On the early models, no provision is made for the external adjustment of the clutch, the operating connecting link being locked with a teper pin, as shown in (!, Fig. 31). Any adjustment necessary to this type of clutch should be done by a competent service station.

Later models, GLG-1 and onwards, are provided with an adjustable link, **D**, as shown in Fig. 32, which is

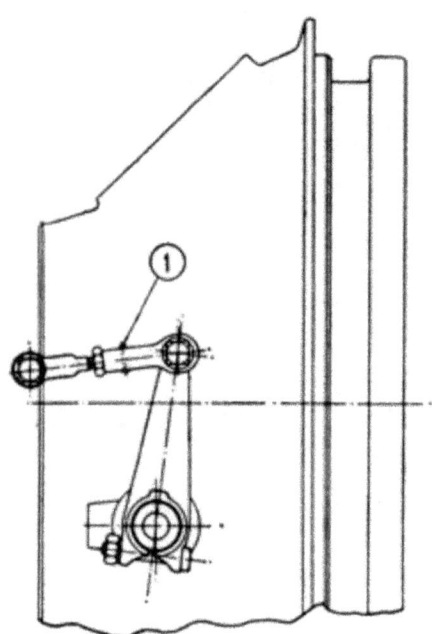

Fig. 31

coupled to the jaw, **E**, by means of a screwed sleeve, **F**, having left and right-hand threads and provided with a hexagon, **F**2.

Release the locknuts, **D**1 and **F**1, and rotate the sleeve, **F**, with a spanner to obtain the correct free movement of the pedal; subsequently retightened the locknuts.

When testing and setting this adjustment, the aluminium pedal plate must be in position,

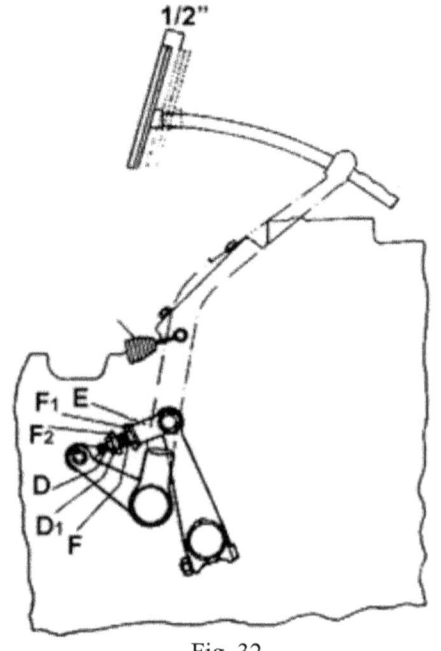

Fig. 32

CHAPTER VI
The Cooling System

Collant Pump - Coolant Level - Frost Precautions.

Coolant Pump.

The centrifugal coolant circulating pump is fitted with a special double packing gland designed to facilitiate lubrication, and tthereby reduce wear.

It is improbable that any leakage or any other trouble will be experienced over long intevals of running, provided that the gland is properly and regularly lubricated.

A screw-down greaser is provided for lubricating the gland and bearings and it should be filled one-third full of grease every 2,500 miles, as directed on page22, and screwed right down, preferably when the engine is warm.

Coolant Level.

This should be inspected daily, and if plain water is being used the level should be maintaned about half way across the upper radiator water pipe, as shown in Fig. **3**.

If an anti-freeze mixture is being used, the level should be maintained so as to just cover the upper tubes of the radiator core.

Frost Precautions.

Where plain water is being used as the coolant medium.and there is any possibility of the car being exposed to low, frosty temperatures, with the engine not running, it is of vital importance that the water system should be drained by opening the drain tap on the water pump and releasing the filler cap. Also, after a frost and before attempting to start, or even move, the engine again, *hot water should first be poured over the water pump*, as otherwise damage may be caused to the pump rotor by the presence of particles of ice within the casing. Warm water can be used with advantage for refilling the radiator.

A suitable anti-freeze mixture is made by mixing soft water with either inhibited ethylene glycol or "Bluecol" in proportions dependent on the degree of frost likely to be encountered.

The following table gives an approximate indication of the amount of frost protection ensured by different strengths of mixture.

Freezing point	22° F.	12° F.	2° F.	-3° F.
Degrees of frost	10° F.	20° F.	30° F.	35° F.
1 Inhibited Ethylene Glycol ...	4½ pts.	6¾ pts.	10 pts	11 pts.
2 "Bluecol"	4½ pts.	6¾ pts.	10pts	11pts.

When changing from water to anti-freeze or from a glycol mixture to a glycerine mixture, the radiator system must be drained. New anti-freeze of the required amount should be mixed with an equal quantity of soft water before being poured into the radiator, the radiator being finally topped up with soft water.

The engine should then be run until normal operating temperature is reached, to ensure uniform distribution of the anti-freeze throughout the system.

The rubber connections must be carefully examined and replaced if unsound, as any leakage will necessitate replenishment with anti-freeze mixture.

When using an anti-freeze mixture as described, a similar mixture should be used for topping up purposes.

THE ELECTRICAL SYSTEM 65

CHAPTER VII

Electrical System

General - Battery - Battery Ignition - Magneto Ignition - Sparking Plugs - Electrical Fault Location.

General.

Four wiring diagrams are illustrated, Figs. **33**, **34**, **35** and **36**, each diagram being appropriate for the chassis series as listed on same.

The electrical system is earthed on the negative side of the battery to the chassis frame.

Battery.

The battery recommended for use on the car may be either of the following: -

Battery Maker's Type Designation.		Voltage	Normal Charging Current
Exide	P. & R. Dagnite.		
6-XSM-1L	6-TBS7-A	12	5 amperes.
6-XHR-1ML	6-BGD9-5		

Never allow the liquid in the cells to fall below the tops of the separators.

Inspect the battery at regular intevals, as directed on page 21, and top up with distilled water, so as to maintain the level of the liquid at ½" above the tops of the plates.

Do not inspect the battery with the aid of a naked light, and on no account disconnect any of the battery terminals or connections when any charge or discharge current is passing, for such a course incurs risk of explosion and involves personal risk.

Battery Ignition.

The battery ignition consists of an ignition coil, **W**, and combined low-tension contact breaker, **X**, and high-tension distributor, **X**I as shown in Fig. **37**.

A ballast resistance, **R**1, is connected in series with the low-tension winding of the coil. Its function is to limit the current taken by the coil at slow speed, or if the ignition switch be accidentally left on while the engine is stopped. It also secures practical equality of intensity of secondary spark at all speeds.

A condensor connected across the contact points is located in a pocket, **X**2, the condensor case and the main body of the contact breaker unit being together in direct electrical connection with the chassis frame.

The insulated terminal of the condensor is connected to the insulated contact, and they are brought out together to the insulated terminal to which the external low-tension connection is made.

In setting the points the gap opening should be .017" to .021".

A few drops of engine oil should be injected into lubricator, **Z** Fig. **37,** every 2,500 miles, as directed on page 22, in order to lubricate the centrifugal ignition timing mechanism. In addition, the oil so injected serves to maintain an oil seal arranged at the base of the ignition tower to protect the contacts from oily vapour from the crankcase, which is liable to cause pitting. At the same time, a trace only of grease should be smeared on the cam surface.

Every 5,000 miles, as directed on page 27, the pivot pin of the low-tension rocker arm should be lubricated by moving aside the retaining spring and putting one drop of oil on the exposed end of the pivot.

The high-tension distributor requires no attention beyond an occasional wiping of the interior with a clean, dry rag.

It is important that the outside of the coil casing should be kept clean. Also, the cover should occasionally be removed and the top of the coil cleaned with a dry rag. Misfiring is some times caused by an accumulation of dirt around the terminals and on the coil casing.

If the timing of the battery ignition should have been deranged, due, for instance, to removal of the cam operating the low-tension rocker, it can be re-set by reference to the flywheel markings which can be seen on removal of the clutch pit cover.

To carry out this operation, the crankshaft should be turned until the mark B.A.I. (battery, advanced ignition) on the flywheel registers with the mark on the casing when No. I piston is commencing its firing stroke. On some models the marks B.L.I. (battery, late ignition) will be found, in place of B.A.I. this should be set as above.

Owing to the fact that on later models a friction-damped spring drive is used for driving the valve gear and all auxilliaries, and that the starting handle operates to turn the crankshaft through the medium of this spring drive, it is important that the crankshaft be rotated for timimg purposes from the *flywheel* end. Also, the starting handle should not have been used at all since the engine was last running.

THE ELECTRICAL SYSTEM

With the ignition lever set fully advanced where the marking is B.A.I., and fully retarded in the case of B.L.I., set the contact breaker cam to be just on the point of causing the contact break (when turning forward) corresponding to No. 1 cylinder.

Fig. 37 - IGNITION COILS, DISTRIBUTOR AND BALLAST RESISTANCE.

A convenient method of determining precisely when the break takes place is by reference to the ammeter. With the ignition switched on, and someone watching the ammeter, the cam should be slowly rotated on the taper of its shaft in the normal direction of rotation until the required peak breaks contact as indicated by the reading of the ammeter. The screw securing the cam should then be tightened.

Magneto Ignition.

The magneto, of a special type, is fitted as a standby, and has no high-tension distributor, but a single high-tension lead, the terminal of which is fitted to the centre of the battery ignition distributor in place of that from the standard ignition when required. The magneto is arranged to be put into service very quickly should the necessity arise, the following operations being performed in the order named:-

1. Remove the battery igniton fuse marked No. 3 from the distribution box, inserting same in the dummy fuse holder in the cover.

Fig. 38 -MAGNETO,

2. Pull out the high-tension terminal **Tb** (Fig. **37**) of the battery ignition from the distributor and replace with the high-tension magneto lead **Tm**, which is carried in a special holder on the ignition tower when not in use. Insert the battery high-tension lead in the holder.
3. Press down the catch **H** (Fig. **38**) projecting from the magneto drive shaft and turn the shaft gently by hand until the teeth are felt to engage.

THE ELECTRICAL SYSTEM

The engine is then ready for running on the magneto, the thunb lever on the switch box being used for switching on and off in the same way as for the battery ignition.

Owing to the fact that the magneto is capable of giving a good spark when retarded, no attempt should be made to start the engine on the magneto ignition, either by hand or by the starter, without first fully retarding the ignition. Also when running on this ignition it will be necessary, inorder to obtain the best results, to use the ignition lever as the engine speed increases or fall off.

When changing back from magneto to battery ignition the operations detailed in the preceding paragraphs 1, 2 and 3, must be reversed, the magnetoe being disconnected by sliding the shaft towards the rear against the pressure of an internal spring until it is felt that the catch is holding the engaging teeth clear of each other, when it will be possible to rotate the shaft by hand.

It is important that this uncoupling of the drive should be effected before running again on the battery ignition.

For chassis series as per Fig. **33** *the following will apply:-*

When re-timimg the magneto reference should be made to the timing markings on the flywheel by removing the clutch pit cover. The crankshaft should be turned until the T.D.C. (top dead centre) mark registers with the timing mark on the flywheel casing. In this position the low-tension contacts should be breaking when the ignition lever is fully retarded.

For chassis series as per Fig. **34**, **35** *and* **36**, *as follows:-*

The magneto timing is set by reference to the mark M.A.I. (magneto, advanced ignition) on the flywheel. When this mark registers with the mark on the casing the low-tension contacts should be just breaking with the ignition lever fully advanced and No. 1 piston approaching the firing position.

Sparking Plugs.

The sparking plugs recommended are KLG, M30 or Champion No. 7 18 m/m..

Every 5,000 miles, as directed on page 28, they should be removed and cleaned. The width of the gaps should be checked and, if necessary, reset to .020".

Electrical Fault Location.

In case of faulty operation, proceed to investigate as follows: -
1. Failure of any part of the system separately may be due to a blown fuse in the distribution box.
2. Failure of incorrect operaton of the system may be due to the fusing of the emergency battery fuse due to an earth
   Repeated failure of a properly fitted fuse indicates a fault on the system.

If the dynamo does not charge: -
1. Brushes stickinng, due probably to oiliness. Clean brushes and holders with rag moistened with petrol.
2. Melting of dynamo armature or field fuse, which latter may be due to:-
    (a) Dirty cut-out contacts, which clean.
    (b) Discontinuity or bad contact in Dynamo battery circuit. See that lights are in order and examine battery terminal connections.
    (c) Sticking dynamo negative brushes.
3. On later series if regulator is fitted:-
    Ascertain whether dynamo or regulator is at fault by removing cover and connecting terminal **I** to **E**, and terminal **A** to **+**. This will short-circuit the regulator. Then, start engine gently and increase speed slowly. If dynamo is in order, the output will be delivered and the defect will lie in the regulator.

If dynamo oputput is low, this may be due to battery being fully charged, but if low with lights on, i.e. ammeter indicates an abnormal discharge, the regulator may be sticking in such a manner as permanently to insert the field resistance.

If dynamo gives an excessive charge and blows fuse when speeded up, this may be due to regulator sticking or to a break in the regulator shunt coil circuit. Check regulator wiring connections.

In the case of defective operation which is traceable to the regulator, the unit must be removed and returned for rectification to Rolls-Royce Ltd.

If, with the fuses intact, and the lights in order, the ignition: -
(a) Misses.
  1. First confirm right condition of sparking plugs.
  2. Assure correct condition of contact breaker points, and adjust gap .017" to .021", if necessary.
  3. If missing still continues, test ignition circuit as below.
(b) Fails.
  1. With battery ignition switched on, see by ammeter, while engine is being cranked, that the coil is taking current intermittently. If no current, test with a small voltmeter (to frame) availability of battery voltage on ballast resistance terminals then at coil terminals.

    If, with battery in order, starter motor is sluggish or does not turn, examine commutator and brushes. Clean oily brushes and holders with a rag moistened with petrol. If motor turns without turning engine, examine Bijur drive.

If battery will not retain charge: -
1. Ascertain that no circuit is left switched on.
2. Test each individual cell with a small voltmeter, with all lights on.
3. See that no cell of the battery leaks acid.

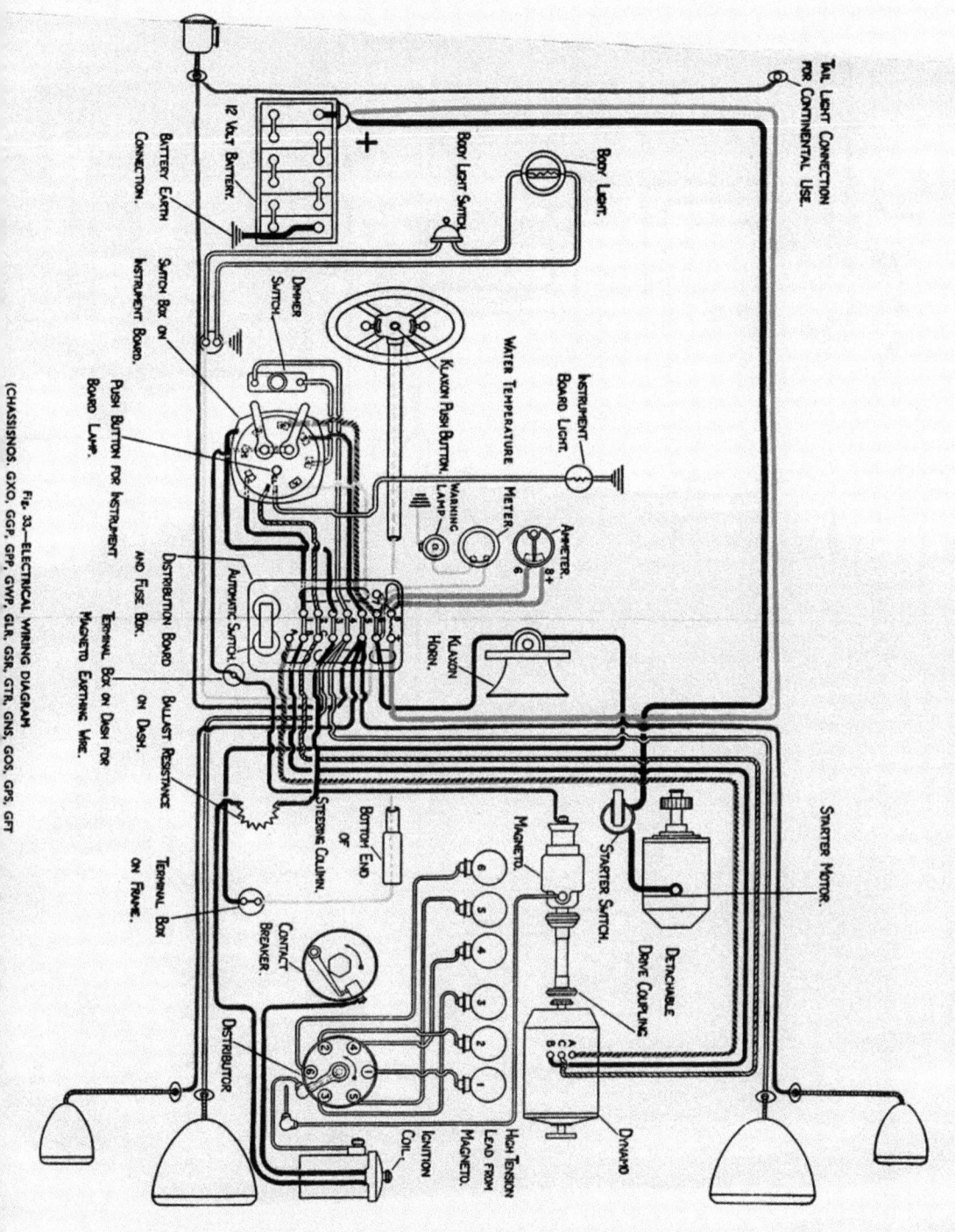

Fig. 33—ELECTRICAL WIRING DIAGRAM
(CHASSIS NOS. GXO, GGP, GPP, GWP, GLR, GSR, GTR, GNS, GOS, GPS, GFT)

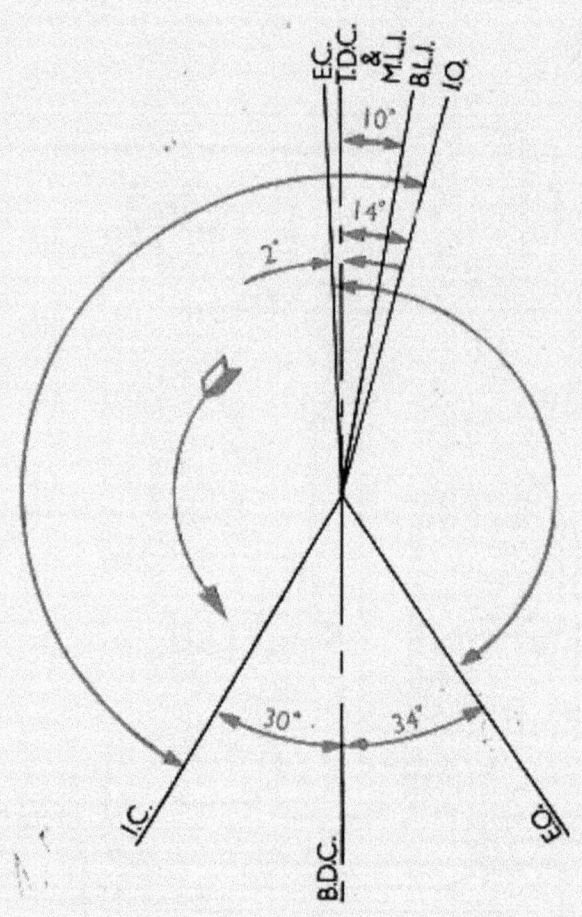

20 H.P. AND 20/25 H.P.

CHASSIS: 112 IN J SERIES TO 24 IN R SERIES

Flywheel Marking Clearance
·020 in. (·051 cm.)
Running Clearance
·004 in. (·010 cm.)
BETWEEN THE VALVE STEM END AND THE ROCKER ARM FACE

www.ingramcontent.com/pod-product-compliance
Ingram Content Group UK Ltd.
Pitfield, Milton Keynes, MK11 3LW, UK
UKHW021432161225
9608UKWH00003B/29